Derived ℓ-Adic Categories for Algebraic Stacks

MEMOIRS
of the
American Mathematical Society

Number 774

Derived ℓ-Adic Categories for
Algebraic Stacks

Kai A. Behrend

May 2003 • Volume 163 • Number 774 (first of 5 numbers) • ISSN 0065-9266

American Mathematical Society
Providence, Rhode Island

2000 *Mathematics Subject Classification.* Primary 14D20, 18Gxx.

Library of Congress Cataloging-in-Publication Data

Behrend, K. (Kai)
 Derived ℓ-adic categories for algebraic stacks / Kai A. Behrend.
  p. cm. — (Memoirs of the American Mathematical Society, ISSN 0065-9266 ; no. 774)
 "Volume 163, number 774 (first of 5 numbers)."
 Includes bibliographical references.
 ISBN 0-8218-2929-7
 1. Algebra, Homological. 2. Moduli theory. 3. Algebraic stacks. I. Title. II. Series.

QA3.A57 no. 774
[QA169]
510 s—dc21
[512′.55]   2003040432

Memoirs of the American Mathematical Society

This journal is devoted entirely to research in pure and applied mathematics.

Subscription information. The 2003 subscription begins with volume 161 and consists of six mailings, each containing one or more numbers. Subscription prices for 2003 are $555 list, $444 institutional member. A late charge of 10% of the subscription price will be imposed on orders received from nonmembers after January 1 of the subscription year. Subscribers outside the United States and India must pay a postage surcharge of $31; subscribers in India must pay a postage surcharge of $43. Expedited delivery to destinations in North America $35; elsewhere $130. Each number may be ordered separately; *please specify number* when ordering an individual number. For prices and titles of recently released numbers, see the New Publications sections of the *Notices of the American Mathematical Society*.

Back number information. For back issues see the *AMS Catalog of Publications*.

Subscriptions and orders should be addressed to the American Mathematical Society, P. O. Box 845904, Boston, MA 02284-5904, USA. *All orders must be accompanied by payment.* Other correspondence should be addressed to 201 Charles Street, Providence, RI 02904-2294, USA.

Copying and reprinting. Individual readers of this publication, and nonprofit libraries acting for them, are permitted to make fair use of the material, such as to copy a chapter for use in teaching or research. Permission is granted to quote brief passages from this publication in reviews, provided the customary acknowledgment of the source is given.

Republication, systematic copying, or multiple reproduction of any material in this publication is permitted only under license from the American Mathematical Society. Requests for such permission should be addressed to the Acquisitions Department, American Mathematical Society, 201 Charles Street, Providence, Rhode Island 02904-2294, USA. Requests can also be made by e-mail to `reprint-permission@ams.org`.

Memoirs of the American Mathematical Society is published bimonthly (each volume consisting usually of more than one number) by the American Mathematical Society at 201 Charles Street, Providence, RI 02904-2294, USA. Periodicals postage paid at Providence, RI. Postmaster: Send address changes to Memoirs, American Mathematical Society, 201 Charles Street, Providence, RI 02904-2294, USA.

© 2003 by the American Mathematical Society. All rights reserved.
This publication is indexed in *Science Citation Index*®, *SciSearch*®, *Research Alert*®, *CompuMath Citation Index*®, *Current Contents*®/*Physical, Chemical & Earth Sciences*.
Printed in the United States of America.

∞ The paper used in this book is acid-free and falls within the guidelines established to ensure permanence and durability.
Visit the AMS home page at `http://www.ams.org/`

10 9 8 7 6 5 4 3 2 1   08 07 06 05 04 03

Contents

Introduction	1
Chapter 1. The ℓ-adic Formalism	7
1. Quotients of t-Categories	7
2. ℓ-Adic Derived Categories	9
2.1. Functoriality	11
Chapter 2. Stratifications	15
1. d-Structures	15
2. cd-Structures	17
Chapter 3. Topoi	21
1. Fibered Topoi	21
2. Constructible Sheaves	24
3. Constructible ℓ-Adic Sheaves	26
4. Topoi with c-Structures and ℓ-Adic Derived Categories	29
5. The d-Structures Defined by c-Topoi	34
6. Topoi with e-Structures	37
Chapter 4. Algebraic Stacks	41
1. Preliminaries on Algebraic Stacks	41
1.1. Gerbe-Like morphisms	41
1.2. Devissage for Algebraic Stacks	43
1.3. Universal Homeomorphisms	45
2. The Étale Topos of an Algebraic Stack	46
2.1. The Flat Topos	46
2.2. The Étale Topos	47
3. The Smooth Topos	48
4. The Simplicial Approach	51
4.1. The d-structures defined by the simplicial c-topos	52
5. The Smooth Approach	53
6. The Étale-Smooth cd-Structure	57
7. The ℓ-Adic d-Structure	62
7.1. Closed L-Stratifications	62
7.2. Some more auxiliary results	64
7.3. The Main Theorem	65
8. Purity and Extraordinary Pullbacks	67

Chapter 5. Convergent Complexes 73
  1. $\overline{Q}_\ell$-Complexes 73
  1.1. A Theorem of Borel 75
  2. Frobenius 75
  3. Mixed and Convergent Complexes 78
  3.1. Convergence 79
  4. The Trace Formula 82
  4.1. More about Frobenius 82
  4.2. Trace of Frobenius 84
  4.3. An Example 89

Bibliography 93

Abstract

We construct an ℓ-adic formalism of derived categories for algebraic stacks. Over finite fields we generalize the theory of mixed complexes to a theory of so called *convergent* complexes. This leads to a proof of the Lefschetz Trace Formula for the Frobenius.

Received by the editor July 25, 1999 and in revised form June 5, 2000.
1991 *Mathematics Subject Classification.* 14D20, 18Gxx.
Key words and phrases. stacks, derived categories, Lefschetz trace formula.
Supported in part by a grant from the National Science Foundation.

Introduction

Motivation. The present paper grew out of the desire to prove the following theorem, conjectured in [3]

THEOREM 0.1 (Lefschetz trace formula for stacks). *Let $\mathfrak{X}$ be a smooth algebraic stack over the finite field $\mathbb{F}_q$. Then*

$$q^{\dim \mathfrak{X}} \operatorname{tr} \Phi_q | H^*(\mathfrak{X}, \mathbb{Q}_\ell) = \#\mathfrak{X}(\mathbb{F}_q). \tag{1}$$

Here Φ_q is the (arithmetic) Frobenius acting on the ℓ-adic cohomology of $\mathfrak{X}$.

For example, if $\mathfrak{X} = B\mathbb{G}_\mathrm{m}$, the classifying stack of the multiplicative group, then $\dim B\mathbb{G}_\mathrm{m} = -1$ (since dividing the 'point' $\operatorname{Spec}\mathbb{F}_q$ of dimension 0 by the 1-dimensional group $\mathbb{G}_\mathrm{m}$ gives a quotient of dimension -1), and $\operatorname{tr}\Phi_q|H^{2i}(B\mathbb{G}_\mathrm{m}, \mathbb{Q}_\ell) = \frac{1}{q^i}$, (since the cohomology of $B\mathbb{G}_\mathrm{m}$ is the cohomology of the infinite-dimensional projective space). Thus the left hand side of (1) is

$$\frac{1}{q}\sum_{i=0}^{\infty}\frac{1}{q^i} = \frac{1}{q-1}.$$

On the right hand side we get the number of $\mathbb{F}_q$-rational points of the stack $\mathfrak{X}$. For the case of $B\mathbb{G}_\mathrm{m}$ this is the number of principal $\mathbb{G}_\mathrm{m}$-bundles over $\operatorname{Spec}\mathbb{F}_q$ (up to isomorphism), or the number of line bundles over $\operatorname{Spec}\mathbb{F}_q$. Since all line bundles over $\operatorname{Spec}\mathbb{F}_q$ are trivial, there is only one isomorphism class in $B\mathbb{G}_\mathrm{m}(\mathbb{F}_q)$. But the number of automorphisms of the trivial line bundle is $\#\mathbb{G}_\mathrm{m}(\mathbb{F}_q) = \#\mathbb{F}_q^* = q-1$, and to count points in the stack sense, we need to divide each point by the number of its automorphisms. Thus the right hand side of (1) is $\frac{1}{q-1}$, also.

As so often in mathematics, it turns out the best way to prove Theorem 0.1 is to generalize it. For example, we wish to stratify our stack $\mathfrak{X}$ and deduce the theorem for $\mathfrak{X}$ from the theorem on the strata. This requires that we consider more general coefficients than $\mathbb{Q}_\ell$ (for example, the sheaves one gets by pushing forward $\mathbb{Q}_\ell$ from the strata). We will also want to perform various base changes, so we consider a relative version of the theorem (for a morphism $\mathfrak{X} \to \mathfrak{Y}$ instead of just a stack $\mathfrak{X}$). This all works very well, if one has a sufficiently flexible *ℓ-adic formalism* at ones disposal. Constructing such a formalism occupied the main part of this paper.

There is a construction of a derived category of equivariant sheaves due to Bernstein and Lunts (see [5]). This is a topological analogue of a special case of our derived category. (The equivariant case is the case of quotient stacks.) The similarity can be seen from their Appendix B, where a description of their category is given, which is formally close to our category.

We construct an ℓ-adic derived category for stacks, called $\mathbb{D}_c^+(\mathfrak{X}, Q_\ell)$, where c stands for 'constructible' (we also have versions where c is replaced by m for 'mixed' and a for '(absolutely) convergent'). We also construct the beginnings of a Grothendieck formalism of six operations, but we restrict ourselves to Rf_*, f^* and $Rf^!$, which are the three operations we need to prove the trace formula.

There are two main problems that one faces if one wants to construct such an ℓ-adic formalism for stacks. First of all, there are the 'topological' problems, consisting in finding the correct site (or topos) with respect to which one defines higher direct images Rf_* and pullbacks f^* and $f^!$. The étale topos is certainly too coarse. The second problem is an ℓ-adic problem dealing with defining a well-behaved ℓ-adic formalism for *unbounded* complexes.

Let us describe these problems in more detail.

Topological Problems. We need a cohomological formalism for constructible torsion sheaves. Let A' be a ring whose characteristic is invertible on all algebraic stacks in question. Let $\mathfrak{X}$ be an algebraic stack. Then we have the category $\mathrm{Mod}(\mathfrak{X}_{\text{ét}}, A')$ of étale sheaves of A'-modules on $\mathfrak{X}$. We wish this category to be the category of coefficient sheaves. Now for algebraic stacks the étale topology is not fine enough to compute the correct cohomology groups of such a sheaf $F \in \mathrm{ob}\,\mathrm{Mod}(\mathfrak{X}_{\text{ét}}, A')$. If, for example, G is a connected algebraic group and $\mathfrak{X} = BG$ is the classifying stack of G, then $BG_{\text{ét}} = S_{\text{ét}}$, where S is the base we are working over (see Corollary 4.30). The stack $B\mathbb{G}_{\text{m}}$, for example, should have the cohomology of the infinite dimensional projective space, but from the étale point of view $B\mathbb{G}_{\text{m}}$ looks like a point. This leads us to consider the smooth topology $\mathfrak{X}_{\text{sm}}$ on the stack $\mathfrak{X}$. We use the smooth topology to compute cohomology of étale coefficient sheaves. Thus we are dealing with a pair of topoi (or sites) $(\mathfrak{X}_{\text{sm}}, \mathfrak{X}_{\text{ét}})$. Since this situation arises in other contexts as well, we formalize it axiomatically. We call such pairs of topoi *c-topoi* (see Section 4).

Thus we consider the derived category $D(\mathfrak{X}_{\text{sm}}, A')$ of the category of smooth sheaves of A'-modules and pass to the subcategory $D_{\text{ét}}(\mathfrak{X}_{\text{sm}}, A')$, defined by requiring the cohomology sheaves to be étale. This gives rise to the definition of $Rf_* : D_{\text{ét}}^+(\mathfrak{X}_{\text{sm}}, A') \to D_{\text{ét}}^+(\mathfrak{Y}_{\text{sm}}, A')$, for a morphism of algebraic stacks $f : \mathfrak{X} \to \mathfrak{Y}$. If $\mathfrak{X} = X$ is a scheme, then we have $D_{\text{ét}}^+(X_{\text{sm}}, A') = D^+(X_{\text{ét}}, A')$ (Proposition 3.42).

One of the problems with the smooth topology is the correct choice of a site defining it. Since all the problems arise already for schemes, let X be a scheme. One possible definition of a smooth site would be to take all smooth X-schemes with smooth morphisms between them. The problem with this definition is that products in this category are not what they should be, if they even exist at all. (For a smooth morphism to have a smooth diagonal it has to be étale.) This means that it is not clear why a morphism $f : X \to Y$

of schemes should even induce a continuous functor f^* of these corresponding sites.

On the other hand, there is the big smooth site. None of the 'topological' difficulties arise in this context, but unfortunately, the direct image functor is not compatible with the étale direct image. It is easy to construct a morphism $f : X \to Y$ of schemes and an étale sheaf F on X, such that f_*F (with respect to the big smooth site) is not étale. So using the big smooth site would not generalize the étale theory for schemes.

So one is forced to use the intermediate notion, where we require objects of the smooth site to smooth over X, but let morphisms be arbitrary. This has the unpleasant side effect that this construction does not commute with localization. For every algebraic stack we get a whole collection of relative smooth sites (see Section 3). A more serious drawback is that, even though f^* is continuous, it is not exact. Thus $f : X \to Y$ does not induce a morphism of the induced smooth topoi, only what we call a *pseudo-morphism* of topoi (see Remark 3.32). Typical counter-examples are closed immersions (see Warning 4.42). So the smooth approach does not give rise to a definition of pullback functors f^*.

This phenomenon necessitates a second approach, the *simplicial* one. We choose a presentation $X \to \mathfrak{X}$ of our algebraic stack $\mathfrak{X}$ which gives rise to a groupoid $X_1 \rightrightarrows X_0$, where $X_0 = X$ and $X_1 = X \times_{\mathfrak{X}} X$. Setting $X_p = \underbrace{X \times_{\mathfrak{X}} \ldots \times_{\mathfrak{X}} X}_{p+1}$ defines a simplicial scheme $X_\bullet$, which has a category of étale sheaves $\text{top}(X_{\bullet\text{ét}})$ associated with it. Again, this gives rise to a c-topos $(\text{top}(X_{\bullet\text{ét}}), \mathfrak{X}_{\text{ét}})$. As above, we consider the derived category $D(X_{\bullet\text{ét}}, A')$ with the corresponding subcategory $D_{\text{cart}}(X_{\bullet\text{ét}}, A')$, defined by requiring the cohomology sheaves to be cartesian. (Note that $\mathfrak{X}_{\text{ét}}$ is the category of cartesian objects of $\text{top}(X_{\bullet\text{ét}})$.) The miracle is that $D_{\text{cart}}^+(X_{\bullet\text{ét}}, A') = D_{\text{ét}}^+(\mathfrak{X}_{\text{sm}}, A')$ (see Proposition 3.63).

So we can use $D_{\text{cart}}^+(X_{\bullet\text{ét}}, A')$ to define f^* and $D_{\text{ét}}^+(\mathfrak{X}_{\text{sm}}, A')$ to define Rf_*. Another miracle is that the functors thus obtained are adjoints of each other. This is due to the fact that there is just enough overlap between the two approaches. The simplicial approach works also for Rf_* for *representable* morphisms, whereas the smooth approach works for f^* if f is *smooth*. These two cases are essentially all that is needed by a devissage lemma (Proposition 4.16).

ℓ-adic Problems. Now let A be a discrete valuation ring (like $\mathbb{Z}_\ell$), whose residue characteristic is invertible on all algebraic stacks we consider. Denote a generator of the maximal ideal of A by ℓ. Then to define an ℓ-adic derived category of *schemes* one defines (see Proposition 2.2.15 of [4]) $D_c^b(X_{\text{ét}}, A) = \text{proj lim}_n D_{\text{ctf}}^b(X_{\text{ét}}, A/\ell^{n+1})$. This approach depends crucially on the fact that the complexes involved are bounded. But in dealing with algebraic stacks we cannot make this convenient restriction. As noted above, the

cohomology of $B\mathbb{G}_m$ is a polynomial ring in one variable, thus represented by an unbounded complex. This is why we have to use a different approach.

(Another ℓ-adic formalism is constructed in [**11**]. But although a triangulated category of unbounded complexes is defined, boundedness is assumed to prove the existence of a t-structure on this triangulated category.)

Overview. We describe our ℓ-adic formalism in Section 1. It uses only results from [**13**, Exp. V]. The main idea is to construct our derived category $\mathbb{D}_c(\mathfrak{X}_{\mathrm{sm}}, A)$ as an inductive limit over the categories $\mathbb{D}_{(\mathcal{S},\mathcal{L})}(\mathfrak{X}_{\mathrm{sm}}, A)$. Here the subscript $(\mathcal{S},\mathcal{L})$ denotes what we call an *L-stratification* (Definition 2.8) of $\mathfrak{X}_{\mathrm{\acute{e}t}}$. Here $\mathcal{S}$ is a stratification of $\mathfrak{X}_{\mathrm{\acute{e}t}}$ and $\mathcal{L}$ associates to every stratum $\mathfrak{V} \in \mathcal{S}$ a finite set $\mathcal{L}(\mathfrak{V})$ of locally constant constructible sheaves of A-modules on $\mathfrak{V}_{\mathrm{\acute{e}t}}$. We require objects of $D_{(\mathcal{S},\mathcal{L})}(\mathfrak{X}_{\mathrm{sm}}, A)$ to have $(\mathcal{S},\mathcal{L})$-constructible cohomology sheaves. The key fact is that the category $\mathrm{Mod}_{(\mathcal{S},\mathcal{L})}(\mathfrak{X}_{\mathrm{\acute{e}t}}, A)$ of $(\mathcal{S},\mathcal{L})$-constructible sheaves of A-modules on $\mathfrak{X}_{\mathrm{\acute{e}t}}$ is finite, i.e. noetherian and artinian. (To be precise, this would be true for $\mathcal{S}$-constructibility, already. The $\mathcal{L}$-part is introduced to deal with higher direct images.)

Section 2 introduces the formalism needed to deal with L-stratifications. We introduce what we call a *d-structure*. This just formalizes the situation of a noetherian topological space X and its locally closed subspaces, where one has the functors i^*, i_*, $i^!$ and $i_!$ between the derived categories on the various locally closed subspaces of X. A *cd-structure* gives the additional data required to introduce L-stratifications.

In Section 3 we apply our ℓ-adic formalism to c-topoi. We reach the central Definition 3.46 of the category of constructible A-complexes on a c-topos X in Section 4. An important role is also played by Proposition 3.59, where d-structures, c-topoi and our ℓ-adic formalism come together, giving rise to an ℓ-adic d-structure on a noetherian c-topos.

In Section 4 we apply our results to algebraic stacks. Our central object of study, $\mathbb{D}_c(\mathfrak{X}_{\mathrm{sm}}, A)$, is introduced in Remark 4.38. The technical heart of this work is contained in Sections 5, 6 and 7. Our main result is summarized in Remark 4.75. In Section 8 we go slightly beyond the two operations f^* and $\mathbb{R}f_*$, by defining $\mathbb{R}f^!$ for certain kinds of representable morphisms of algebraic stacks (essentially smooth morphisms and closed immersions).

Finally, in Section 5, we consider the case of finite ground field $\mathbb{F}_q$. Our goal is to generalize the triangulated category $\mathbb{D}_m^b(X_{\mathrm{\acute{e}t}}, Q_\ell)$ of bounded mixed complexes on a scheme X to the case of algebraic stacks. We introduce (Definition 5.16) the category $\mathbb{D}_m^+(\mathfrak{X}_{\mathrm{sm}}, Q_\ell)$, which is obtained form $\mathbb{D}_c^+(\mathfrak{X}_{\mathrm{sm}}, Q_\ell)$ by requiring cohomology objects to be mixed. This concept behaves well with respect to the three operations f^*, $\mathbb{R}f_*$ and $\mathbb{R}f^!$.

But to define the trace of Frobenius on mixed objects we need a further restriction. We pass to $\mathbb{D}_a^+(\mathfrak{X}_{\mathrm{sm}}, Q_\ell)$, the subcategory of $\mathbb{D}_m^+(\mathfrak{X}_{\mathrm{sm}}, Q_\ell)$ consisting of *absolutely convergent* objects. Roughly speaking, an object $M \in \mathrm{ob}\,\mathbb{D}_m^+(\mathfrak{X}_{\mathrm{sm}}, Q_\ell)$ is absolutely convergent, if for every finite field $\mathbb{F}_{q^n}$

and every morphism $x : \operatorname{Spec} \mathbb{F}_{q^n} \to \mathfrak{X}$, the trace of the arithmetic Frobenius $\sum_i (-1)^i \operatorname{tr} \Phi_q | h^i(\mathbb{R}x^! M)$ is absolutely convergent, no matter how we embed $\overline{Q}_\ell$ into $\mathbb{C}$. We show that the triangulated categories $\mathbb{D}_a^+(\mathfrak{X}_{\mathrm{sm}}, A)$ are stable under the two operations $\mathbb{R}f_*$ and $\mathbb{R}f^!$. The question of stability under f^* remains open.

In the final section (Section 4) we show that our formalism of convergent complexes is rich enough to prove the general Lefschetz Trace Formula for the arithmetic Frobenius on algebraic stacks. Our main result is given in Definition 5.33 and Theorem 5.38. As Corollary 5.39 we get the result we conjectured in [**3**]. To finish, we give a rather interesting example, due to P. Deligne. We show how our trace formula applied to the stack $\mathfrak{M}_1$ of curves of genus one, may be interpreted as a type of Selberg Trace Formula. It gives the sum $\sum_k \frac{1}{p^{k+1}} \operatorname{tr} T_p | \mathcal{S}_{k+2}$, where T_p is the pth Hecke operator on the space of cusp forms of weight $k+2$, in terms of elliptic curves over the finite field $\mathbb{F}_p$.

Notations and Conventions. Our reference for algebraic stacks is [**16**]. We always assume all algebraic stacks to be locally noetherian, in particular quasi-separated. *Stacks* will usually be denoted by German letters $\mathfrak{X}$, $\mathfrak{Y}$ etc., whereas for *spaces* we use Roman letters X, Y etc. A gerbe $\mathfrak{X}/X$ is called *neutral*, if it has a section, i.e. if it is isomorphic to $B(G/X)$, for some relative algebraic space of groups G/X. For a group scheme G, we denote by G° its connected component of the identity. Presentations $X \to \mathfrak{X}$ of an algebraic stack $\mathfrak{X}$ are always smooth, and of finite type if $\mathfrak{X}$ is of finite type. For an algebraic stack $\mathfrak{X}$ we denote by $|\mathfrak{X}|$ the set of isomorphism classes of points of $\mathfrak{X}$. We consider $|\mathfrak{X}|$ as a topological space with respect to the Zariski topology.

If I is a category, we denote by $\operatorname{ob} I$ the objects, by $\operatorname{fl} I$ the arrows in I and by I_{op} the dual category. A box in a commutative diagram denotes a cartesian diagram.

The natural numbers start with zero. $\mathbb{N} = \{0, 1, 2, \ldots\}$.

Acknowledgment. I would like to thank Prof. P. Deligne for encouraging me to undertake this work and for pointing out the example of Section 4 to me.

Kai Behrend

CHAPTER 1

The ℓ-adic Formalism

1. Quotients of t-Categories

For the definition of t-categories we refer to [**4**, 1.3]. Roughly speaking, a t-category is a triangulated category $\mathcal{D}$, having, for every $i \in \mathbb{Z}$, truncation functors $\tau_{\leq i}$ and $\tau_{\geq i}$ and cohomology functors $h^i : \mathcal{D} \to \mathcal{C}$, where $\mathcal{C}$ is the *heart* of $\mathcal{D}$, which is an abelian subcategory of $\mathcal{D}$. We will discuss methods of constructing sub- and quotient-t-categories of a given t-category $\mathcal{D}$.

Let $\mathcal{D}$ be a t-category with heart $\mathcal{C}$. Let $h^i : \mathcal{D} \to \mathcal{C}$ denote the associated cohomology functors.

First, we will construct t-subcategories of $\mathcal{D}$.

LEMMA 1.1. *Let $\mathcal{C}'$ be a full non-empty subcategory of $\mathcal{C}$. Then the following are equivalent.*

(1) *The category $\mathcal{C}'$ is abelian and closed under extensions in $\mathcal{C}$. The inclusion functor $\mathcal{C}' \to \mathcal{C}$ is exact.*
(2) *The category $\mathcal{C}'$ is closed under kernels cokernels and extensions in $\mathcal{C}$.*
(3) *If*
$$A_1 \xrightarrow{f} A_2 \longrightarrow A_3 \longrightarrow A_4 \xrightarrow{g} A_5$$
is an exact sequence in $\mathcal{C}$ and A_1, A_2, A_4 and A_5 are in $\mathcal{C}'$, then so is A_3.

PROOF. Considering the exact sequence
$$0 \longrightarrow \operatorname{cok} f \longrightarrow A_3 \longrightarrow \ker g \longrightarrow 0$$
this is immediate. □

DEFINITION 1.2. We call a subcategory $\mathcal{C}'$ of $\mathcal{C}$ *closed* if it is a full non-empty subcategory satisfying any of the conditions of Lemma 1.1.

Now let $\mathcal{C}'$ be a closed subcategory of $\mathcal{C}$. We may define the full subcategory $\mathcal{D}'$ of $\mathcal{D}$ by requiring an object M of $\mathcal{D}$ to be in $\mathcal{D}'$, if for every $i \in \mathbb{Z}$ the object $h^i M$ of $\mathcal{C}$ is contained in $\mathcal{C}'$. Then using the above lemma it is an easy exercise to prove that $\mathcal{D}'$ is a triangulated subcategory of $\mathcal{D}$. One defines a triangle in $\mathcal{D}'$ to be distinguished if it is distinguished as a triangle in $\mathcal{D}$. It is just as immediate (see also [**4**, 1.3.19]) that we get an induced t-structure on $\mathcal{D}'$ as follows. If $(\mathcal{D}^{\leq 0}, \mathcal{D}^{\geq 0})$ is the t-structure on $\mathcal{D}$, then $(\mathcal{D}' \cap \mathcal{D}^{\leq 0}, \mathcal{D}' \cap \mathcal{D}^{\geq 0})$ is the induced t-structure on $\mathcal{D}'$. The truncation and cohomology functors are obtained by restricting from $\mathcal{D}$ to $\mathcal{D}'$. The heart of $\mathcal{D}'$ is $\mathcal{D}' \cap \mathcal{C}$, which is equal to $\mathcal{C}'$. So we have proved

PROPOSITION 1.3. *The category $\mathcal{D}'$ is a t-category with heart $\mathcal{C}'$. The inclusion functor $\mathcal{D}' \to \mathcal{D}$ is exact. If $\mathcal{D}$ is non-degenerate, then so is $\mathcal{D}'$.*

We now consider the case of quotients of $\mathcal{D}$. For this construction we need a thick subcategory $\mathcal{C}'$ of $\mathcal{C}$. This means that $\mathcal{C}'$ is a non-empty full subcategory satisfying the following condition. Whenever
$$0 \longrightarrow A' \longrightarrow A \longrightarrow A'' \longrightarrow 0$$
is a short exact sequence in $\mathcal{C}$, then A is in $\mathcal{C}'$ if and only if A' and A'' are. Note that this condition implies that $\mathcal{C}'$ is closed in $\mathcal{C}$. Thus the above construction can be carried out and we get a t-subcategory $\mathcal{D}'$ of $\mathcal{D}$ with heart $\mathcal{C}'$. Note that $\mathcal{D}'$ satisfies the following condition. If $M \to N$ is a morphism in $\mathcal{D}$ whose cone is in $\mathcal{D}'$ and which factors through an object of $\mathcal{D}'$, then M and N are in $\mathcal{D}'$. In other words, $\mathcal{D}'$ is a thick subcategory of $\mathcal{D}$.

PROPOSITION 1.4. *The quotient category $\mathcal{D}/\mathcal{D}'$ is in a natural way a t-category with heart $\mathcal{C}/\mathcal{C}'$. The quotient functor $\mathcal{D} \to \mathcal{D}/\mathcal{D}'$ is exact. If $\mathcal{D}$ is non-degenerate, then so is $\mathcal{D}/\mathcal{D}'$.*

PROOF. We define a morphism $A \to B$ in $\mathcal{C}$ to be an *isomorphism modulo $\mathcal{C}'$* if its kernel and cokernel are in $\mathcal{C}'$. The category $\mathcal{C}/\mathcal{C}'$ is endowed with a functor $\mathcal{C} \to \mathcal{C}/\mathcal{C}'$ that turns every isomorphism modulo $\mathcal{C}'$ into an isomorphism. Moreover, $\mathcal{C}/\mathcal{C}'$ is universal with this property.

Call a morphism $M \to N$ in $\mathcal{D}$ a *quasi-isomorphism modulo $\mathcal{C}'$* if for every $i \in \mathbb{Z}$ the homomorphism $h^i M \to h^i N$ is an isomorphism modulo $\mathcal{C}'$. Note that $M \to N$ is a quasi-isomorphism modulo $\mathcal{C}'$ if and only if the cone of $M \to N$ is in $\mathcal{D}'$. This proves that the quasi-isomorphisms modulo $\mathcal{C}'$ form a multiplicative system and the category $\mathcal{D}/\mathcal{D}'$ is the universal category turning the quasi-isomorphisms modulo $\mathcal{C}'$ into isomorphisms. It is standard (see for example [**14**]) that $\mathcal{D}/\mathcal{D}'$ is a triangulated category if we call a triangle in $\mathcal{D}/\mathcal{D}'$ distinguished if it is isomorphic to the image of a distinguished triangle in $\mathcal{D}$.

If $(\mathcal{D}^{\leq 0}, \mathcal{D}^{\geq 0})$ defines the t-structure on $\mathcal{D}$, then the essential images of $\mathcal{D}^{\leq 0}$ and $\mathcal{D}^{\geq 0}$ in $\mathcal{D}/\mathcal{D}'$ define a t-structure on $\mathcal{D}/\mathcal{D}'$ whose associated truncation and cohomology functors are obtained from the corresponding functors for $\mathcal{D}$ through factorization. It is not difficult to prove that the inclusion $\mathcal{C} \to \mathcal{D}$ factors to give a functor $\mathcal{C}/\mathcal{C}' \to \mathcal{D}/\mathcal{D}'$ that identifies $\mathcal{C}/\mathcal{C}'$ as the heart of the t-structure on $\mathcal{D}/\mathcal{D}'$. □

REMARK 1.5. Let $\mathcal{D}_1$ and $\mathcal{D}_2$ be t-categories with hearts $\mathcal{C}_1$ and $\mathcal{C}_2$, respectively. Let $\mathcal{C}'_1$ and $\mathcal{C}'_2$ be thick subcategories of $\mathcal{C}_1$ and $\mathcal{C}_2$, giving rise to thick subcategories $\mathcal{D}'_1$ and $\mathcal{D}'_2$ as above. Let $F : \mathcal{D}_1 \to \mathcal{D}_2$ be an exact functor. If F maps $\mathcal{D}'_1$ to $\mathcal{D}'_2$, then F induces a functor $\widetilde{F} : \mathcal{D}_1/\mathcal{D}'_1 \to \mathcal{D}_2/\mathcal{D}'_2$.

If F has a left adjoint $G : \mathcal{D}_2 \to \mathcal{D}_1$ mapping $\mathcal{D}'_2$ to $\mathcal{D}'_1$, then $\widetilde{G} : \mathcal{D}_2/\mathcal{D}'_2 \to \mathcal{D}_1/\mathcal{D}'_1$ is a left adjoint of $\widetilde{F}$.

REMARK 1.6. Clearly, the constructions of Propositions 1.3 and 1.4 commute with passage to $\mathcal{D}^+$, $\mathcal{D}^-$ or $\mathcal{D}^b$.

2. ℓ-Adic Derived Categories

Using results from [**13**, Exp. V] we will develop an ℓ-adic formalism as follows.

Let A be a discrete valuation ring and $\mathfrak{l}$ the maximal ideal of A. Let ℓ be a generator of $\mathfrak{l}$. For $n \in \mathbb{N}$ we denote $A/\mathfrak{l}^{n+1}$ by Λ_n. For an abelian A-category $\mathfrak{A}$ we denote by $\mathfrak{A}^n$ the category of $(\Lambda_n, \mathfrak{A})$-modules. In other words, $\mathfrak{A}^n$ is the full subcategory of $\mathfrak{A}$ consisting of those objects on which ℓ^{n+1} acts as zero. By $\widetilde{\mathfrak{A}}$ we denote the category of projective systems $(M_n)_{n \in \mathbb{N}}$ in $\mathfrak{A}$ such that $M_n \in \mathfrak{A}^n$ for every $n \in \mathbb{N}$. The categories $\widetilde{\mathfrak{A}}$ and $\mathfrak{A}^n$ are abelian. The nth component functor $a_n : \widetilde{\mathfrak{A}} \to \mathfrak{A}^n$ is exact.

By construction, for $m \geq n$, the category $\mathfrak{A}^n$ is naturally a subcategory of $\mathfrak{A}^m$. This inclusion has a left adjoint $\mathfrak{A}^m \to \mathfrak{A}^n$, which we denote by $M \mapsto M \otimes \Lambda_n$. We may define $M \otimes \Lambda_n = M/\ell^{n+1}M$.

We say that an object $(M_n)_{n \in \mathbb{N}}$ of $\widetilde{\mathfrak{A}}$ is ℓ-adic, if for every $m \geq n$ the morphism $M_m \otimes \Lambda_n \to M_n$ (which is given by adjunction) is an isomorphism. We denote the full subcategory of ℓ-adic projective systems in $\mathfrak{A}$ by $\widetilde{\mathfrak{A}}_\ell$.

An object $M = (M_n)_{n \in \mathbb{N}}$ of $\widetilde{\mathfrak{A}}$ is called AR-null if there exists an integer r such that $M_{n+r} \to M_n$ is the zero map for every $n \in \mathbb{N}$. We say that a homomorphism $\phi : M \to N$ in $\widetilde{\mathfrak{A}}$ is an AR-isomorphism if $\ker \phi$ and $\cok \phi$ are AR-null. An object M of $\widetilde{\mathfrak{A}}$ is called AR-ℓ-adic, if there exists an ℓ-adic object F and an AR-isomorphism $\phi : F \to M$. The symbol $\widetilde{\mathfrak{A}}_{\text{AR-}\ell}$ denotes the category of AR-ℓ-adic objects in $\widetilde{\mathfrak{A}}$.

Some important properties of ℓ-adic objects with respect to AR-isomorphisms are the following:

LEMMA 1.7. *Let $\alpha : M' \to M$ be an AR-isomorphism in $\widetilde{\mathfrak{A}}$, where M is ℓ-adic. Then α has a section.*

LEMMA 1.8. *Let $\alpha : F \to M$ be an AR-isomorphism in $\widetilde{\mathfrak{A}}$, with F being ℓ-adic. Then there exists an $r \geq 0$ such that F_n is a direct summand of $M_{n+r} \otimes \Lambda_n$ for every n.*

PROOF. This follows from the fact that an AR-isomorphism gives rise to an isomorphism in the category of projective systems modulo translation. □

Now let $\mathfrak{A}_c \subset \mathfrak{A}$ be a finite subcategory, closed in $\mathfrak{A}$. This means (Definition 1.2) that $\mathfrak{A}_c$ is a full subcategory closed under kernels, cokernels and extensions in $\mathfrak{A}$, such that every object of $\mathfrak{A}_c$ is noetherian and artinian. For $n \in \mathbb{N}$, we let $\mathfrak{A}_c^n$ be the intersection of $\mathfrak{A}^n$ and $\mathfrak{A}_c$. Moreover, $\widetilde{\mathfrak{A}}_c$ will denote the full subcategory of $\widetilde{\mathfrak{A}}$ consisting of those projective systems each of whose components is in $\mathfrak{A}_c$. It is clear that $\widetilde{\mathfrak{A}}_c$ is closed in $\widetilde{\mathfrak{A}}$. The symbols $\widetilde{\mathfrak{A}}_{\ell,c}$ and $\widetilde{\mathfrak{A}}_{\text{AR-}\ell,c}$ denote the categories of ℓ-adic and AR-ℓ-adic systems in $\widetilde{\mathfrak{A}}_c$, respectively.

LEMMA 1.9. *Let $\alpha : F \to M$ be an AR-isomorphism in $\widetilde{\mathfrak{A}}$, where F is ℓ-adic. If M is in $\widetilde{\mathfrak{A}}_c$, then so is F.*

PROOF. Choose r as in Lemma 1.8. Then if M_{n+r} is in $\mathfrak{A}_c$, so is $M_{n+r} \otimes \Lambda_n$, as the cokernel of multiplication by ℓ^{n+1}. Then F_n is also in $\mathfrak{A}_c$ as a direct summand. □

PROPOSITION 1.10. *The category $\widetilde{\mathfrak{A}}_{AR\text{-}\ell,c}$ is closed in $\widetilde{\mathfrak{A}}$. In particular, it is an abelian category.*

PROOF. Let $\alpha : M \to N$ be a homomorphism in the category $\widetilde{\mathfrak{A}}_{\text{AR-}\ell,c}$. Then $\ker \alpha$ and $\text{cok}\, \alpha$ are in $\widetilde{\mathfrak{A}}_c$. We would like to show that they are AR-ℓ-adic. But this follows from Proposition 5.2.1 of [**13**, Exp. V] applied to the category $\mathcal{C} = \mathfrak{A}_c$.

Now let M and N be objects of $\widetilde{\mathfrak{A}}_{\text{AR-}\ell,c}$ and let
$$0 \longrightarrow M \longrightarrow E \longrightarrow N \longrightarrow 0$$
be an extension of N by M in $\widetilde{\mathfrak{A}}$. Since E is then also in $\widetilde{\mathfrak{A}}_c$, applying Corollaire 5.2.5 of [loc. cit.] to $\mathcal{C} = \mathfrak{A}_c$, we see that E is in fact AR-ℓ-adic. □

We may now carry out the construction of Proposition 1.3, using the subcategory $\widetilde{\mathfrak{A}}_{\text{AR-}\ell,c}$ of $\widetilde{\mathfrak{A}}$. We get a sub-t-category of the derived category $D(\widetilde{\mathfrak{A}})$, which we shall denote by $D_{\text{AR-}\ell,c}(\widetilde{\mathfrak{A}})$.

We shall now construct a certain quotient of $D_{\text{AR-}\ell,c}(\widetilde{\mathfrak{A}})$.

PROPOSITION 1.11. *The objects of $\widetilde{\mathfrak{A}}_{AR\text{-}\ell,c}$ that are AR-null form a thick subcategory.*

PROOF. This is clear. See also [**13**, Exp. V, Proposition 2.2.2]. □

Thus we may apply Proposition 1.4 to our situation. We denote the quotient of $\widetilde{\mathfrak{A}}_{\text{AR-}\ell,c}$ modulo the AR-null sheaves by AR-$\widetilde{\mathfrak{A}}_{\text{AR-}\ell,c}$. Thus AR-$\widetilde{\mathfrak{A}}_{\text{AR-}\ell,c}$ is obtained by inverting the AR-isomorphisms in $\widetilde{\mathfrak{A}}_{\text{AR-}\ell,c}$. We call a morphism $M \to N$ in $D_{\text{AR-}\ell,c}(\widetilde{\mathfrak{A}})$ a *quasi-AR-isomorphism* if the induced map $h^i M \to h^i N$ is an AR-isomorphism for every $i \in \mathbb{Z}$. The triangulated category obtained from $D_{\text{AR-}\ell,c}(\widetilde{\mathfrak{A}})$ by inverting the quasi-AR-isomorphisms will be denoted by AR-$D_{\text{AR-}\ell,c}(\widetilde{\mathfrak{A}})$.

PROPOSITION 1.12. *The natural functor $\widetilde{\mathfrak{A}}_{\ell,c} \to \text{AR-}\widetilde{\mathfrak{A}}_{AR\text{-}\ell,c}$ is an equivalence of categories.*

PROOF. This is easily proved using Lemma 1.7 and Lemma 1.9. □

2. ℓ-ADIC DERIVED CATEGORIES

We will use the functor of Proposition 1.12 to identify the two categories $\widetilde{\mathfrak{A}}_{\ell,c}$ and AR-$\widetilde{\mathfrak{A}}_{\text{AR-}\ell,c}$. For brevity we will denote the t-category AR-$D_{\text{AR-}\ell,c}(\widetilde{\mathfrak{A}})$ by $\mathbb{D}_c(\mathfrak{A})$. Thus $\mathbb{D}_c(\mathfrak{A})$ is an A-t-category with heart $\widetilde{\mathfrak{A}}_{\ell,c}$.

As an example, let us assume that A has finite residue field and let us consider the category $\mathfrak{A} = \text{Mod}(A)$ of A-modules. Then $\text{Mod}(A)^n$ is the category of Λ_n-modules. As finite subcategory, closed in $\text{Mod}(A)$ we take the category $\text{Mod}_c(A)$ of finite A-modules. By $\widetilde{\text{Mod}_{\ell,c}(A)}$ we denote the category of ℓ-adic systems of finite A-modules.

PROPOSITION 1.13. *Let $\hat{A}$ be the ℓ-adic completion of A. There is an equivalence of categories*

$$\widetilde{\text{Mod}_{\ell,c}(A)} \longrightarrow \text{Mod}_{fg}(\hat{A}),$$

where $\text{Mod}_{fg}(\hat{A})$ is the category of finitely generated $\hat{A}$-modules.

PROOF. We may define this equivalence by $(M_n)_n \mapsto \text{proj}\lim_n M_n$. An inverse is given by $M \mapsto (M \otimes \Lambda_n)_n$. □

In this situation, we denote the t-category $\mathbb{D}_c(\text{Mod}(A))$ by $\mathbb{D}_c(A)$. So $\mathbb{D}_c(A)$ is a t-category with heart $\text{Mod}_{fg}(\hat{A})$.

PROPOSITION 1.14. *The category $\mathbb{D}_c(A)$ is naturally equivalent to $D_{fg}(\hat{A})$, the subcategory of the derived category of $\text{Mod}(\hat{A})$ consisting of complexes whose cohomology is finitely generated over $\hat{A}$.*

PROOF. This is a straightforward exercise deriving the two functors $(M_n)_n \mapsto \text{proj}\lim_n M_n$ and $M \mapsto (M \otimes \Lambda_n)_n$ between the categories $\widetilde{\text{Mod}(A)}$ and $\text{Mod}(\hat{A})$. □

2.1. Functoriality. Let $\mathfrak{A}$ and $\mathfrak{B}$ be abelian A-categories and let $F : \mathfrak{A} \to \mathfrak{B}$ be an A-linear functor. Then for every $n \in \mathbb{N}$ there is an induced functor $F^n : \mathfrak{A}^n \to \mathfrak{B}^n$ and thus an induced functor $\widetilde{F} : \widetilde{\mathfrak{A}} \to \widetilde{\mathfrak{B}}$. Any exactness properties of F carry over to F^n and $\widetilde{F}$.

If $\mathfrak{A}$ has sufficiently many injectives, then so does $\mathfrak{A}^n$, for every n. This is easily seen by noticing that if I is an injective object of $\mathfrak{A}$, then the kernel of the action of ℓ^{n+1} on I is injective in $\mathfrak{A}^n$.

LEMMA 1.15. *If $\mathfrak{A}$ has sufficient injectives then so does $\widetilde{\mathfrak{A}}$.*

PROOF. Choose for every $n \in \mathbb{N}$ an injective object I_n of $\mathfrak{A}^n$. Then define $J_n = I_1 \times \ldots \times I_n$ and $J_m \to J_n$ for $m \geq n$ to be the projection onto the first n factors. Then $(J_n)_{n \in \mathbb{N}}$ is an injective object of $\widetilde{\mathfrak{A}}$, and every object of $\widetilde{\mathfrak{A}}$ admits an injection into an object of this type. □

Assume now that $\mathfrak{A}$ has enough injectives and that F is left exact. Then the right derived functor $RF : D^+(\mathfrak{A}) \to D^+(\mathfrak{B})$ exists. By the above remarks, we also have the existence of the right derived functors $RF^n : D^+(\mathfrak{A}^n) \to D^+(\mathfrak{B}^n)$ and $R\widetilde{F} : D^+(\widetilde{\mathfrak{A}}) \to D^+(\widetilde{\mathfrak{B}})$. Let us assume that

RF commutes with restriction of scalars, i.e. that for every $n \in \mathbb{N}$ the commutative diagram

$$\begin{array}{ccc} \mathfrak{A}^n & \longrightarrow & \mathfrak{A} \\ F^n \downarrow & & \downarrow F \\ \mathfrak{B}^n & \longrightarrow & \mathfrak{B} \end{array}$$

derives to give a commutative diagram

$$\begin{array}{ccc} D^+(\mathfrak{A}^n) & \longrightarrow & D^+(\mathfrak{A}) \\ RF^n \downarrow & & \downarrow RF \\ D^+(\mathfrak{B}^n) & \longrightarrow & D^+(\mathfrak{B}). \end{array}$$

It is easily seen that we then also get commutative diagrams

$$\begin{array}{ccc} D^+(\widetilde{\mathfrak{A}}) & \xrightarrow{a_n} & D^+(\mathfrak{A}^n) \\ R\widetilde{F} \downarrow & & RF^n \downarrow \\ D^+(\widetilde{\mathfrak{B}}) & \xrightarrow{b_n} & D^+(\mathfrak{B}^n), \end{array}$$

for every $n \in \mathbb{N}$.

Now assume furthermore that $\mathfrak{A}_c$ and $\mathfrak{B}_c$ are finite closed subcategories of $\mathfrak{A}$ and $\mathfrak{B}$. The following result is essentially proved in [**13**, Exp. V].

LEMMA 1.16. *If the derived functors $R^i F : \mathfrak{A} \to \mathfrak{B}$ map $\mathfrak{A}_c$ into $\mathfrak{B}_c$, then $R\widetilde{F}$ maps $D^+_{AR\text{-}\ell,c}(\widetilde{\mathfrak{A}})$ into $D^+_{AR\text{-}\ell,c}(\widetilde{\mathfrak{B}})$.*

PROOF. First note that for any object M of $D^+(\widetilde{\mathfrak{A}})$ we have an E_2-spectral sequence

$$R^i \widetilde{F}(h^j M) \Longrightarrow h^{i+j}(R\widetilde{F}(M)) \qquad (2)$$

in the abelian category $\widetilde{\mathfrak{B}}$. So it suffices to prove that if $M = (M_n)_{n \in \mathbb{N}}$ is an object of $\widetilde{\mathfrak{A}}_{\text{AR-}\ell,c}$, then the higher direct images $R^i \widetilde{F} M$ are in $\widetilde{\mathfrak{B}}_{\text{AR-}\ell,c}$. But we have $(R^i \widetilde{F} M)_n = R^i F M_n$ by the above commutative diagrams. By assumption, $R^i F M_n$ is in $\mathfrak{B}_c$.

We also have an exact A-linear δ-functor $(R^i F)_i$ from $\mathfrak{A}_c$ to $\mathfrak{B}_c$. If R is a finitely generated graded Λ_0-algebra then any functor from $\mathfrak{A}_c$ to $\mathfrak{B}_c$ takes the category of graded noetherian $(R, \mathfrak{A}_c)$-modules into the category of graded noetherian $(R, \mathfrak{B}_c)$-modules. By this fact Proposition 5.3.1 of [**13**, Exp. V] applies to the δ-functor $(R^i F) : \mathfrak{A}_c \to \mathfrak{B}_c$. Thus $(R^i F M_n)_n$ is AR-ℓ-adic. $\square$

COROLLARY 1.17. *Assume that the derived functors $R^i F : \mathfrak{A} \to \mathfrak{B}$ map $\mathfrak{A}_c$ into $\mathfrak{B}_c$. Then we get an induced left t-exact functor $\mathbb{R}F : \mathbb{D}^+_c(\mathfrak{A}) \to \mathbb{D}^+_c(\mathfrak{B})$. For $i \geq 0$ let us denote the induced functor $h^i \mathbb{R}F : \mathfrak{A}_{\ell,c} \to \mathfrak{B}_{\ell,c}$ by $\mathbb{R}^i F$. We have for every i a commutative diagram*

$$\begin{array}{ccc} \widetilde{\mathfrak{A}}_{AR\text{-}\ell,c} & \xrightarrow{R^i \widetilde{F}} & \widetilde{\mathfrak{B}}_{AR\text{-}\ell,c} \\ \downarrow & & \downarrow \\ \widetilde{\mathfrak{A}}_{\ell,c} & \xrightarrow{\mathbb{R}^i F} & \widetilde{\mathfrak{B}}_{\ell,c}. \end{array}$$

For every $M \in \text{ob}\,\mathbb{D}_c^+(\mathfrak{A})$ there is an E_2-spectral sequence
$$\mathbb{R}^i F(h^j M) \Longrightarrow h^{i+j}\mathbb{R}F(M)$$
in the abelian category $\mathfrak{B}_{\ell,c}$.

PROOF. First we need to show that if $M \to N$ is a quasi-AR-isomorphism in $D^+_{\text{AR-}\ell,c}(\widetilde{\mathfrak{A}})$ then $R\widetilde{F}M \to R\widetilde{F}N$ is quasi-AR-isomorphism in $D^+_{\text{AR-}\ell,c}(\widetilde{\mathfrak{B}})$. Using the spectral sequence (2) we reduce to proving that if M is an AR-zero object of $\widetilde{\mathfrak{A}}$, then $R^i\widetilde{F}(M)$ is AR-zero, for all $i \geq 0$. But this is clear. The other claims are now also clear. □

CHAPTER 2

Stratifications

The following concepts are introduced in [**4**], although without giving them names.

1. d-Structures

For the basic definitions and constructions concerning fibered categories see [**12**, Exp. VI]. Let us fix a ring A. For a topological space X, let $\mathrm{lc}(X)$ denote the set of locally closed subsets of X, considered as a category with respect to inclusion. For an object of $\mathrm{lc}(X)$, we sometimes denote the morphism $V \to X$ by j_V.

DEFINITION 2.1. Let X be a noetherian topological space. A *stratification* $\mathcal{S}$ *of* X is a finite number of locally closed nonempty subsets of X, called the *strata*, such that X is the disjoint union of $\mathcal{S}$, and such that the closure of each stratum is the union of strata.

Note that this definition implies that each stratum is open it its closure. Note also that every decomposition of X into finitely many locally closed subsets may be refined to a stratification. The pullback of a stratification by a continuous map is a stratification.

DEFINITION 2.2. Let X be a noetherian topological space. An *A-d-structure on* X (or just a *d-structure*) is a category D, fibered over $\mathrm{lc}(X)$, such that

(1) For every $V \in \mathrm{ob}\,\mathrm{lc}(X)$ the fiber $D(V)$ is an A-t-category and $D(\varnothing) = 0$. All pullback functors are exact and A-linear.
(2) If $i : V \to W$ is a closed immersion in $\mathrm{lc}(X)$, then any pullback functor $i^* : D(W) \to D(V)$ has an exact A-linear right adjoint i_*, which has an exact A-linear right adjoint $i^!$. The functor i_* is fully faithful and t-exact.
(3) If $j : V \to W$ is an open immersion in $\mathrm{lc}(X)$, then any pullback functor $j^* : D(W) \to D(V)$ has an exact A-linear left adjoint $j_!$ and an exact A-linear right adjoint j_*. The functors $j_!$ and j_* are fully faithful and j^* is t-exact.
(4) If $V \in \mathrm{ob}\,\mathrm{lc}(X)$ is the disjoint union of U and Z, where $j : U \to V$ is an open immersion and $i : Z \to V$ is a closed immersion, then $j^* i_* = 0$. For any $M \in \mathrm{ob}\,D(V)$ there exist homomorphisms

$i_*i^*M \to j_!j^*M[1]$ and $j_*j^*M \to i_*i^!M[1]$ such that

$$\begin{array}{ccc}
 & i_*i^*M & \\
 \nearrow & & \nwarrow \\
j_!j^*M & \longrightarrow & M
\end{array}$$

and

$$\begin{array}{ccc}
 & j_*j^*M & \\
 \nearrow & & \nwarrow \\
i_*i^!M & \longrightarrow & M
\end{array}$$

are distinguished triangles in $D(V)$.

For some elementary facts implied by these axioms, see Section 1.4 of [4]. Note that every locally closed subspace of X is naturally endowed with an induced d-structure. Whenever $k : V \to W$ is a morphism in $\mathrm{lc}(X)$, we have naturally two pairs of adjoint functors (k^*, k_*) and $(k_!, k^!)$ between $D(V)$ and $D(W)$. If k is a closed immersion we have $k_* = k_!$, if k is an open immersion we have $k^* = k^!$.

REMARK 2.3. Of course, we may also define a d-structure on X as a fibered category over $\mathrm{lc}(X)_{\mathrm{op}}$, the dual category of $\mathrm{lc}(X)$, using the pushforward functors to define the fibered structure. In this case, we denote for $k : V \to W$ in $\mathrm{fl\,lc}(X)$ the pullback functor by k_*. It is clear how the above axioms can be adapted to this case.

Reversing all arrows, we return to the category $\mathrm{lc}(X)$, but pass to the dual category of each fiber $D(V)_{\mathrm{op}}$. Then the pushforward functors k_* define a co-fibered category over $\mathrm{lc}(X)$. Then the existence of the left adjoints k^* (which are now actually right adjoints) shows that $V \mapsto D(V)_{\mathrm{op}}$ is also fibered over $\mathrm{lc}(X)$. In other words, we have a bi-fibered category $V \mapsto D(V)_{\mathrm{op}}$ over $\mathrm{lc}(X)$. Of course, the above axioms may also be adapted to this viewpoint.

Let $f_0 : X \to Y$ be a continuous map of noetherian topological spaces with A-d-structures D_X and D_Y. The map f_0 induces a functor $f_0^{-1} : \mathrm{lc}(Y) \to \mathrm{lc}(X)$ preserving open and closed immersions. Via f_0^{-1} we may pull back the d-structure D_X on X and get a category $(f_0^{-1})^*D_X$, fibered over $\mathrm{lc}(Y)$.

DEFINITION 2.4. A *morphism* $f : D_X \to D_Y$ *of d-structures* (covering f_0) is a cartesian $\mathrm{lc}(Y)$-functor $f^* : D_Y \to (f_0^{-1})^*D_X$, such that
  (1) For every $V \in \mathrm{ob\,lc}(Y)$ the fiber functor $f^* : D(V) \to D(f_0^{-1}V)$ is A-linear, right t-exact and admits an exact A-linear right adjoint $f_* : D(f_0^{-1}V) \to D(V)$.
  (2) If $j : V \to W$ is an open immersion in $\mathrm{lc}(Y)$ and $M \in \mathrm{ob\,}D(f_0^{-1}W)$ then the natural homomorphism $j^*f_*M \to f_*j^*M$ in $D(V)$ is an isomorphism.

Whenever $k : V \to W$ is a morphism in $\mathrm{lc}(X)$, then k induces in an obvious way a morphism of d-structures $k : D_V \to D_W$, where D_V and D_W are the d-structures induced by D on V and W, respectively. Moreover, if $f : X \to Y$ is a morphism of d-structures, and $V \in \mathrm{ob}\,\mathrm{lc}(X)$ and $W \in \mathrm{ob}\,\mathrm{lc}(Y)$ are such that $f_0(V) \subset W$, then we can construct an induced morphism of d-structures $f : V \to W$, where V and W are endowed with the induced d-structures.

2. cd-Structures

DEFINITION 2.5. Let X be a noetherian topological space. A *cd-structure on X* is a d-structure D on X endowed with a full fibered subcategory D_{lcc}, such that for every $V \in \mathrm{ob}\,\mathrm{lc}(X)$ the subcategory $D_{\mathrm{lcc}}(V)$ of $D(V)$ satisfies

(1) Let $C(V)$ be the heart of $D(V)$ and $C_{\mathrm{lcc}}(V) = D_{\mathrm{lcc}} \cap C(V)$. Then $C_{\mathrm{lcc}}(V)$ is a finite closed subcategory of $C(V)$.
(2) If $M \in \mathrm{ob}\,D(V)$ and for all $i \in \mathbb{Z}$ we have $h^i M \in \mathrm{ob}\,C_{\mathrm{lcc}}(V)$, then $M \in \mathrm{ob}\,D_{\mathrm{lcc}}(V)$.
(3) We have $D_{\mathrm{lcc}}(V) = D_{\mathrm{lcc}}^+(V)$, i.e. for every $M \in \mathrm{ob}\,D_{\mathrm{lcc}}(V)$ it holds that $h^i M = 0$, for all $i \mathbb{L} 0$.

If X is a noetherian topological space with a cd-structure $D_{\mathrm{lcc}} \subset D$, then every locally closed subspace of X naturally inherits an induced cd-structure.

DEFINITION 2.6. A *pre-L-stratification* of X is a pair $(\mathcal{S}, \mathcal{L})$, where $\mathcal{S}$ is a stratification of X and $\mathcal{L}$ assigns to every stratum $V \in \mathcal{S}$ a finite set $\mathcal{L}(V)$ of simple objects of $C_{\mathrm{lcc}}(V)$.

An object $M \in \mathrm{ob}\,D(X)$ is called $(\mathcal{S}, \mathcal{L})$-*constructible*, if for every $V \in \mathcal{S}$ and every $i \in \mathbb{Z}$ we have that $h^i(j_V^* M)$ is an object of $C_{\mathrm{lcc}}(V)$, whose Jordan-Hölder components are isomorphic to elements of $\mathcal{L}(V)$.

An object $M \in \mathrm{ob}\,D(X)$ is called *constructible*, if there exists a pre-L-stratification $(\mathcal{S}, \mathcal{L})$ of X such that M is $(\mathcal{S}, \mathcal{L})$-constructible. In this case we also say that the pre-L-stratification $(\mathcal{S}, \mathcal{L})$ trivializes M.

Let $D_c(X)$ be the full subcategory of $D(X)$ consisting of constructible objects. Clearly, $D_c(X)$ is an A-t-category.

DEFINITION 2.7. The cd-structure $D_{\mathrm{lcc}} \subset D$ on X is called *tractable*, if for any open immersion $j : V \to W$ in $\mathrm{lc}(X)$ any pushforward $j_* : D(V) \to D(W)$ takes constructible objects to constructible objects.

Note that for every object V of $\mathrm{lc}(X)$ the induced cd-structure on V is tractable, if the given cd-structure on X is tractable.

DEFINITION 2.8. Let $D_{\mathrm{lcc}} \subset D$ be a tractable cd-structure on X. We call a pre-L-stratification $(\mathcal{S}, \mathcal{L})$ an *L-stratification*, if for any $V \in \mathcal{S}$ and any $L \in \mathcal{L}(V)$ the pushforward $j_{V*}L$ is $(\mathcal{S}, \mathcal{L})$-constructible.

For the following considerations, let us fix a tractable cd-structure on X.

DEFINITION 2.9. Let $(\mathcal{S}, \mathcal{L})$ and $(\mathcal{S}', \mathcal{L}')$ be pre-L-stratifications of X. We say that $(\mathcal{S}', \mathcal{L}')$ is a *refinement* of $(\mathcal{S}, \mathcal{L})$ if $\mathcal{S}'$ is a refinement of $\mathcal{S}$ and for every stratum $V \in \mathcal{S}$, every $L \in \mathcal{L}(V)$ is trivialized by $(\mathcal{S}', \mathcal{L}')|V$.

LEMMA 2.10. *Every pre-L-stratification may be refined to an L-stratification.*

PROOF. Let $(\mathcal{S}, \mathcal{L})$ be a pre-L-stratification of X. The set of open subsets U of X such that $(\mathcal{S}, \mathcal{L})|U$ can be refined to an L-stratification is non-empty. This follows by considering an open stratum. Since X is noetherian, there exists thus a maximal open subset U of X such that $(\mathcal{S}, \mathcal{L})$ can be refined to an L-stratification over U. So let $(\mathcal{S}', \mathcal{L}')$ be an L-stratification of U refining $(\mathcal{S}, \mathcal{L})|U$. Let $Z = X - U$. If $Z \neq \emptyset$ there exists a non-empty open subset $V \subset Z$ such that

(1) $j_{W*}L|V$ is lcc for every inclusion j_W of a stratum W of $\mathcal{S}'$ into X and every L in $\mathcal{L}'(W)$.
(2) If $W \subset U$ is a stratum of $\mathcal{S}'$, then the closure $\overline{W}$ of W in X either contains V or does not intersect V.
(3) V is contained in a stratum of $\mathcal{S}$.

Then $U \cup V$ is an open subset of X over which $(\mathcal{S}, \mathcal{L})$ can be refined to an L-stratification. This contradiction implies that $U = X$. □

LEMMA 2.11. *Let $(\mathcal{S}, \mathcal{L})$ be an L-stratification. Let $k : V \to W$ be a morphism in $\mathrm{lc}(X)$, where both V and W are the union of strata from $\mathcal{S}$. Then k^*, k_*, $k^!$ and $k_!$ preserve $(\mathcal{S}, \mathcal{L})$-constructability.*

PROOF. Let us first deal with k_*. We easily reduce to the case that $W = X$. If V consists of more than one stratum we may write V as a disjoint union $V = U \cap Z$, where $i : Z \to V$ is a closed and $j : U \to V$ an open immersion. Using induction on the number of strata V consists of, we may assume that our claim holds for $(ki)_*$, $(kj)_*$ and j_*. If $M \in \mathrm{ob}\, D(V)$ is $(\mathcal{S}, \mathcal{L})$-constructible, then so is j_*j^*M and hence $i_*i^!M$ and $i^*i_*i^!M = i^!M$. Thus $(ki)_*i^!M$ and $(kj)_*j^*M$ are $(\mathcal{S}, \mathcal{L})$-constructible, which implies that k_*M is $(\mathcal{S}, \mathcal{L})$-constructible, which is what we wanted to prove. It remains to treat the case that V consists of one stratum. So let $M \in \mathrm{ob}\, D(V)$ be $(\mathcal{S}, \mathcal{L})$-constructible. Since $h^iM = 0$ for $i \gg 0$ we reduce to the case that $M \in \mathrm{ob}\, C_{\mathrm{lcc}}(V)$. An easy induction on the length of M finishes the proof.

The case of $k^!$ follows from the case of k_*. The cases $k_!$ and k^* are trivial. □

LEMMA 2.12. *Let $(\mathcal{S}, \mathcal{L})$ be an L-stratification. An object $M \in \mathrm{ob}\, D(X)$ is $(\mathcal{S}, \mathcal{L})$-constructible, if and only if for every $V \in \mathcal{S}$ and every $i \in \mathbb{Z}$ we have that $h^i(j_V^! M)$ is an object of $C_{\mathrm{lcc}}(V)$, whose Jordan-Hölder components are isomorphic to elements of $\mathcal{L}(V)$.*

PROOF. The 'only if' part was proved in Lemma 2.11. So assume now that $h^i j_V^! M$ has Jordan-Hölder components in $\mathcal{L}(V)$ for all $V \in \mathcal{S}$ and $i \in \mathbb{Z}$. Let us prove that M is $(\mathcal{S}, \mathcal{L})$-constructible by induction on the number of strata in $\mathcal{S}$. The case of one stratum being trivial, let us assume that $\mathcal{S}$ contains more than one stratum. Thus we may decompose X into a disjoint union $X = U \cup Z$, where U and Z are unions of strata in $\mathcal{S}$ and $i : Z \to X$ is a closed and $j : U \to X$ an open immersion. We may assume that the lemma holds for U and Z. Hence $i^! M$ and $j^* M$ are $(\mathcal{S}, \mathcal{L})$-constructible. By Lemma 2.11, $i_* i^! M$ and $j_* j^* M$ are $(\mathcal{S}, \mathcal{L})$-constructible, hence M is. □

DEFINITION 2.13. Let $f : X \to Y$ be a morphism of d-structures D_X and D_Y. Assume that D_X and D_Y are endowed with cd-structures $D_{\text{lcc},X}$ and $D_{\text{lcc},Y}$. The morphism f is called a *morphism of cd-structures*, if the cartesian $\text{lc}(Y)$-functor $f^* : D_Y \to (f_0^{-1})^* D_X$ maps $D_{\text{lcc},Y}$ to $(f_0^{-1})^* D_{\text{lcc},X}$.

DEFINITION 2.14. Let X and Y be endowed with tractable cd-structures and let $f : X \to Y$ be a morphism of cd-structures. We say that f is *tractable*, if any pushforward functor $f_* : D(X) \to D(Y)$ maps constructible objects to constructible objects.

For example, any morphism $k : V \to W$ in $\text{lc}(X)$, where X is a cd-structure, gives rise to a morphism of the induced cd-structures in a natural manner. If X is tractable, then the morphism of cd-structures $k : V \to W$ is tractable.

Let $f : X \to Y$ be a morphism of cd-structures. If $V \in \text{ob lc}(X)$ and $W \in \text{ob lc}(Y)$ are such that $f_0(V) \subset W$, then the induced morphism of d-structures $f : V \to W$ is a morphism of cd-structures. Now assume that $f : X \to Y$ is tractable (which implies by definition that X and Y are tractable). Let $W \in \text{ob lc}(Y)$. Then any pushforward functor $f_* : D(f_0^{-1} W) \to D(W)$ maps constructible objects to constructible objects. This is because we may write $f_* M = j_W^! f_* j_{f_0^{-1} W *} M$, for $M \in \text{ob } D(f_0^{-1} W)$. In particular, if we have $V \in \text{ob lc}(X)$ and $W \in \text{ob lc}(Y)$ such that $f_0(V) \subset W$, then the induced morphism of cd-structures $f : V \to W$ is tractable.

Obviously, a composition of tractable morphisms of cd-structures is tractable.

LEMMA 2.15. *Let $f : X \to Y$ be a morphism of tractable cd-structures. Let X be the disjoint union of a closed subset Z and an open subset U. If $f|U$ and $f|Z$ are tractable, then so is f.*

PROOF. Let $M \in \text{ob } D(X)$. We have a distinguished triangle in $D(Y)$

$$(f|Z)_* i^! M \swarrow \quad (f|U)_* j^* M \quad \searrow \longrightarrow f_* M \qquad (3)$$

which immediately implies the result. □

LEMMA 2.16. *Let $f : X \to Y$ be a tractable morphism of cd-structures. If $(\mathcal{S}, \mathcal{L})$ is an L-stratification of X, then there exists an L-stratification $(\mathcal{S}', \mathcal{L}')$ of Y such that whenever $M \in \operatorname{ob} D(X)$ is $(\mathcal{S}, \mathcal{L})$-constructible, the pushforward $f_* M$ is $(\mathcal{S}', \mathcal{L}')$-constructible.*

PROOF. Consider for every $V \in \mathcal{S}$ and every $L \in \mathcal{L}(V)$ the constructible object $j_{V*} L$ of $D(X)$. Take $(\mathcal{S}', \mathcal{L}')$ to be an L-stratification of Y trivializing all $f_* j_{V*} L$, which is possible, since these are finite in number. Now let $M \in \operatorname{ob} D(X)$ be $(\mathcal{S}, \mathcal{L})$-constructible. We claim that $f_* M$ is $(\mathcal{S}', \mathcal{L}')$-constructible. Decomposing X into a closed Z and an open U, such that $\mathcal{S}$ is a refinement of the stratification $\{U, Z\}$ of X, we get a distinguished triangle (3) in $D(Y)$. Using induction on the number of strata in $\mathcal{S}$, this allows us to reduce to the case that $\mathcal{S}$ contains only one stratum. Using the fact that $h^i M = 0$ for $i \ll 0$ we reduce to the case that $M \in \operatorname{ob} C_{\operatorname{lcc}}(X)$ and another induction on the length of M finishes the proof. $\square$

CHAPTER 3

Topoi

1. Fibered Topoi

Recall the following facts about fibered topoi (see Section 7 of [**2**, Exp. VI]). Let I be a category. A fibered topos $X_\bullet$ over I is a fibered category $X_\bullet$ over I, such that for every $n \in \operatorname{ob} I$ the fiber category X_n is a topos and for every $\alpha : n \to m$ in $\operatorname{fl}(I)$ any pullback functor $\alpha^* : X_m \to X_n$ is the pullback functor of a morphism of topoi $\alpha : X_n \to X_m$.

For example, if X is a given topos and $\phi : I \to X$ is a functor, then we get a corresponding fibered topos $X_\bullet$ by setting $X_n = X_{/\phi(n)}$, the induced topos over $\phi(n)$.

If $X_\bullet$ is a fibered topos over I, let us denote by $\operatorname{top}(X_\bullet)$ the associated total topos (Remarque 7.4.3 of [loc. cit.]). Having chosen pullback functors for $X_\bullet$, we may describe $\operatorname{top}(X_\bullet)$ as the category whose objects are families $F^\bullet = ((F^n)_{n \in \operatorname{ob} I}, (\theta(\alpha))_{\alpha \in \operatorname{fl}(I)})$, where $F^n \in \operatorname{ob} X_n$ and for $\alpha : n \to m$, $\theta(\alpha)$ is a morphism $\theta(\alpha) : \alpha^* F^m \to F^n$ in X_n. This data is subject to the obvious cocycle condition, namely that $\theta(\alpha \circ \beta) = \theta(\beta) \circ \beta^* \theta(\alpha)$. A morphism $g^\bullet : F^\bullet \to G^\bullet$ in $X_\bullet$ is a family $g^\bullet = (g^n)_{n \in \operatorname{ob} I}$, where $g^n : F^n \to G^n$ is a morphism in X_n, subject to the obvious commuting relations, namely that $g^n \circ \theta_F(\alpha) = \theta_G(\alpha) \circ \alpha^*(g^m)$, for any $\alpha : n \to m$ in $\operatorname{fl}(I)$. Composition in $\operatorname{top}(X_\bullet)$ is defined in the obvious manner.

If no confusion seems likely to arise, we write $X_\bullet$ for $\operatorname{top}(X_\bullet)$.

Denote for $n \in \operatorname{ob} I$ the functor $\operatorname{top}(X_\bullet) \to X_n$, which assigns to $F^\bullet \in \operatorname{ob} \operatorname{top}(X_\bullet)$ its component F^n over n, by ι_n^*. Then ι_n^* has a left adjoint $\iota_{n!}$ and a right adjoint ι_{n*}, which are fully faithful if n has no non-trivial endomorphisms in $\operatorname{fl}(I)$. Thus we have a morphism of topoi $\iota_n : X_n \to \operatorname{top}(X_\bullet)$. Explicitly, we have for an object $F \in \operatorname{ob} X_n$:

$$\iota_m^* \iota_{n!} F = \coprod_{\alpha \in \operatorname{Hom}(m,n)} \alpha^* F$$

$$\iota_m^* \iota_{n*} F = \prod_{\alpha \in \operatorname{Hom}(n,m)} \alpha_* F.$$

By the exactness of ι_n^*, in $\operatorname{top}(X_\bullet)$ fibered products, disjoint sums and quotients by equivalence relations may be calculated componentwise. Also, a family in $\operatorname{top}(X_\bullet)$ is covering, if and only if it is covering componentwise.

Let $X_\bullet$ be a fibered topos over I. Let $X_{\bullet\operatorname{cart}}$ be the category of cartesian sections of $X_\bullet$. The category $X_{\bullet\operatorname{cart}}$ is the subcategory of $\operatorname{top}(X_\bullet)$ consisting of those sheaves $F^\bullet$, for which all transition morphisms $\theta(\alpha): \alpha^* F^m \to F^n$ are isomorphisms. The above remarks show that $X_{\bullet\text{cart}}$ is, in fact, a topos. Denoting the inclusion functor into $\text{top}(X_\bullet)$ by $\pi^*: X_{\bullet\text{cart}} \to \text{top}(X_\bullet)$, we see that π^* has a right adjoint π_* such that $\pi_*\pi^* = \text{id}$. So we get a morphism of topoi $\pi: \text{top}(X_\bullet) \to X_{\bullet\text{cart}}$.

EXAMPLE 3.1. Let Δ be the category of standard simplices, whose set of objects we denote by $\{\Delta_0, \Delta_1, \ldots\}$. A topos fibered over Δ_{op} is called a *simplicial topos*.

Let X be a topos and U an object of X, covering the final object. Define the functor $\phi: \Delta_{\text{op}} \to X$ by $\phi(\Delta_n) = U^{n+1}$ and by taking face maps to projections and degeneracy maps to diagonals, so as to obtain a simplicial object $U_\bullet$ in X, the *Čech nerve* of the one-element covering U of X. Denote by $U_\bullet$ also the corresponding fibered topos over Δ_{op}. Clearly, we have $U_{\bullet\text{cart}} = X$ and so we have for every $n \in \mathbb{N}$ a commutative diagram of topoi

$$\begin{array}{ccc} U_n & \xrightarrow{\iota_n} & \text{top}(U_\bullet) \\ & {}_{j_n}\searrow & \downarrow \pi \\ & & X, \end{array}$$

where we have written U_n for U_{Δ_n}, and $j_n: U_n \to X$ for the localization morphism.

LEMMA 3.2. *With the notation of Example 3.1 we have that every abelian cartesian sheaf on $\text{top}(U_\bullet)$ is acyclic for π_*.*

PROOF. This is Čech cohomology. $\square$

EXAMPLE 3.3. Consider $\mathbb{N}$ as a category,

$$\mathbb{N} = \{0 \to 1 \to 2 \to \ldots\},$$

and let X be a topos. Consider the fibered topos $X_\bullet$ over $\mathbb{N}$ given by the constant functor $\mathbb{N} \to X$ mapping to the final object in X. The the corresponding total topos $X^\mathbb{N} = \text{top}(X_\bullet)$ is just the category of projective systems in X. A cartesian object in $X^\mathbb{N}$ is a constant projective system. Thus the pullback morphism of the morphism of topoi $\pi: X^\mathbb{N} \to X$ is the associated constant projective system functor. Note that $\iota_{0*} = \pi^*$. Moreover, $\pi_* = \text{proj lim}$.

More generally, If $X_\bullet$ is a fibered topos over I, then we get an induced fibered topos $X_\bullet^\mathbb{N}$ over I. We have $\text{top}(X_\bullet^\mathbb{N}) = \text{top}(X_\bullet)^\mathbb{N}$ and $(X_\bullet^\mathbb{N})_{\text{cart}} = (X_{\bullet\text{cart}})^\mathbb{N}$.

Given a discrete valuation ring A with parameter ℓ, we consider the projective system $\Lambda = (\Lambda_n)_{n \in \mathbb{N}}$, where $\Lambda_n = A/\ell^{n+1}$, which is a sheaf of rings on $X^\mathbb{N}$. Note that an object M of $\text{Mod}(X^\mathbb{N}, \Lambda)$ is just a projective system $M = (M_n)_{n \in \mathbb{N}}$ of sheaves of A-modules on X, satisfying $\ell^{n+1} M_n = 0$ for every $n \in \mathbb{N}$. If we denote our A-category $\text{Mod}(X, A)$ by $\mathfrak{A}$ then in the notation of Section 2 we have $\mathfrak{A}^n = \text{Mod}(X, \Lambda_n)$ and $\widetilde{\mathfrak{A}} = \text{Mod}(X^\mathbb{N}, \Lambda)$. We denote the full subcategory of $\mathrm{Mod}(X^{\mathbb{N}}, \Lambda)$ consisting of ℓ-adic objects by $\mathrm{Mod}_\ell(X^{\mathbb{N}}, \Lambda)$. We call $\mathrm{Mod}_\ell(X^{\mathbb{N}}, \Lambda)$ the *category of ℓ-adic sheaves on X*.

Let $X_\bullet$ and $Y_\bullet$ be fibered topoi over I. A cartesian morphism of fibered topoi $f_\bullet : X_\bullet \to Y_\bullet$ is a cartesian I-functor $f^*_\bullet : Y_\bullet \to X_\bullet$, such that for every $n \in \mathrm{ob}\, I$ the fiber functor $f_n^* : Y_n \to X_n$ is the pullback functor of a morphism of topoi $f_n : X_n \to Y_n$. A cartesian morphism of fibered topoi gives rise to a morphism $f : \mathrm{top}(X_\bullet) \to \mathrm{top}(Y_\bullet)$. The derived functors Rf_* may be calculated componentwise. This follows from the fact that a flasque sheaf on $\mathrm{top}(X_\bullet)$ induces a flasque sheaf in each component, as is proved in the proof of Lemme 8.7.2 of [**2**, Exp. VI].

Let $f : X_\bullet \to Y_\bullet$ be a cartesian morphism of fibered topoi. We call f an immersion (open immersion, closed immersion, locally closed immersion) if $f_n : X_n \to Y_n$ is one, for all $n \in \mathrm{ob}\, I$. Moreover, we call an immersion $f : X_\bullet \to Y_\bullet$ *strict*, if for every $\alpha : n \to m$ in $\mathrm{fl}\, I$ the commutative diagram of topoi

$$\begin{array}{ccc} X_n & \xrightarrow{f_n} & Y_n \\ \alpha \downarrow & & \downarrow \alpha \\ X_m & \xrightarrow{f_m} & Y_m \end{array}$$

is 2-cartesian.

Note that if f is an immersion then so is the induced morphism $f : \mathrm{top}(X_\bullet) \to \mathrm{top}(Y_\bullet)$. This is easily seen using the descriptions of $\mathrm{top}(X_\bullet)$ and $\mathrm{top}(Y_\bullet)$ in terms of α_*-morphisms.

DEFINITION 3.4. Let $X_\bullet$ be a fibered topos over I. A *fibered subtopos* $V_\bullet$ of $X_\bullet$ is given by a subtopos V_n of X_n for every $n \in \mathrm{ob}\, I$ such that for every $\alpha : n \to m$ of $\mathrm{fl}\, I$ and any object F of V_n we have that $\alpha_* F$ is contained in V_m. We say that $V_\bullet$ is an *open (closed, locally closed)* fibered subtopos, if for every $n \in \mathrm{ob}\, I$ the subtopos V_n is open (closed, locally closed) in X_n. Moreover, we call $V_\bullet$ a *strict* subtopos of $X_\bullet$ if for every $\alpha : n \to m$ in $\mathrm{fl}\, I$ we have that $\alpha^{-1}(V_m) = V_n$. Here $\alpha^{-1}(V_m)$ denotes the inverse image of V_m in X_n (see Exercice 9.1.6 in [**2**, Exp. IV]).

Let $V_\bullet$ be a fibered subtopos of $X_\bullet$.

NOTE 3.5. Choose functors α_* for $X_\bullet$. Then we get induced functors $\alpha_* : V_n \to V_m$ which are direct image functors of morphisms of topoi $\alpha : V_n \to V_m$, by Proposition 9.1.3 in [**2**, Exp. IV]. In this way, we may make $V_\bullet$ into a fibered topos over I together with an immersion $V_\bullet \to X_\bullet$ of fibered topoi.

WARNING 3.6. Let $\alpha : n \to m$ be in $\mathrm{fl}\, I$. Then $\alpha^* : X_m \to X_n$ does *not*, in general, map V_m to V_n. So we may *not* think of $V_\bullet$ as a fibered subcategory of $X_\bullet$. Already for open immersions this does not work, although for closed immersions it does.

DEFINITION 3.7. Let $V_\bullet \subset X_\bullet$ be a fibered subtopos. We get an induced immersion $\text{top}(V_\bullet) \to \text{top}(X_\bullet)$. The essential image of its direct image functor is called the *total subtopos of* $\text{top}(X_\bullet)$ *defined by* $V_\bullet \subset X_\bullet$. By abuse of notation, we will also denote it by $\text{top}(V_\bullet)$.

PROPOSITION 3.8. *Let* $U_\bullet$ *be a strict open subtopos of* $X_\bullet$. *Let us denote for every* $n \in \text{ob}\, I$ *the closed complement of* U_n *by* Z_n. *Then* $Z_\bullet$ *is a strict closed subtopos of* $X_\bullet$. *Moreover,* $j : \text{top}(U_\bullet) \to \text{top}(X_\bullet)$ *is an open immersion with closed complement* $i : \text{top}(Z_\bullet) \to \text{top}(X_\bullet)$.

The functors $j_!$, j^*, Rj_* *and* i^*, i_*, $Ri^!$ *between* $D^+(X_\bullet, \Lambda^\bullet)$, $D^+(Z_\bullet, i^*\Lambda^\bullet)$ *and* $D^+(U_\bullet, j^*\Lambda^\bullet)$ *may be calculated componentwise, for any sheaf of rings* $\Lambda^\bullet = (\Lambda^n)_{n \in \text{ob}\, I}$ *on* $\text{top}(X_\bullet)$.

PROPOSITION 3.9. *Let* $f : X_\bullet \to Y_\bullet$ *be a cartesian morphism of fibered topoi and* $V_\bullet \subset Y_\bullet$ *a locally closed fibered subtopos. Let* $W_n = f_n^{-1}(V_n)$, *for every* $n \in \text{ob}\, I$. *Then* $W_\bullet$ *is a locally closed fibered subtopos of* $X_\bullet$. *We write* $W_\bullet = f^{-1}(V_\bullet)$ *and call it the pullback of* $V_\bullet$.

If $V_\bullet \subset Y_\bullet$ *is a strict locally closed fibered subtopos then* $W_\bullet$ *is a strict locally closed subtopos of* $X_\bullet$ *and we have* $\text{top}(W_\bullet) = f^{-1}(\text{top}(V_\bullet))$.

COROLLARY 3.10. *Let* $X_\bullet$ *be a fibered topos and* V *a locally closed subtopos of* $X_{\bullet\, cart}$. *Let* $V_\bullet$ *be the (strict) fibered subtopos of* $X_\bullet$ *given by* $V_n = (\pi \circ \iota_n)^{-1} V$, *for every* $n \in \text{ob}\, I$. *Then* $\text{top}(V_\bullet) = \pi^{-1} V$.

2. Constructible Sheaves

Let X be a topos. We will always assume that X has sufficiently many points. By $|X|$ we will denote a conservative set of points, for example a set of representatives for the collection of all isomorphism classes of points of X. We will think of $|X|$ as a topological space (see Exercice 7.8(a) in [**2**, Exp. IV]). Recall (Exercice 9.7.5 of [**2**, Exp. IV]) that the locally closed subtopoi of X are in bijective correspondence to the locally closed subspaces of $|X|$.

In addition, let us assume that $|X|$ is noetherian. Fix a ring A.

DEFINITION 3.11. For any locally closed subtopos V of X we let $D(V, A)$ be the derived category of $\text{Mod}(V, A)$, the category of sheaves of A-modules on V. Then the collection of the various $D^+(V, A)$ is naturally a d-structure (see Definition 2.2) on $|X|$. Let us call $D^+(\,\cdot\,, A)$ the *canonical* A-*d-structure on* X.

Note that any morphism of topoi induces a morphism of the associated canonical A-d-structures.

DEFINITION 3.12. A sheaf F on X is called an *lcc sheaf* if locally F is isomorphic to the constant sheaf associated to a finite set. A morphism $F \to G$ in X is called an *lcc morphism* if it makes F into an lcc object of the induced topos $X_{/G}$.

PROPOSITION 3.13. *The following are some basic facts about lcc morphisms.*

(1) *Any base change of an lcc morphism is lcc.*
(2) *The property of being lcc is local on the base.*
(3) *Let $f = g \circ h$ be a composition of morphisms in X. If g and h are lcc, then so is f. If f and g are lcc, then so is h. If h is an epimorphism and f and h are lcc, then so is g.*
(4) *Fibered products of lcc sheaves are lcc.*
(5) *Let G be a group sheaf in X and $E \to B$ a principal G-bundle in X, such that B covers X. Then E and B are lcc if and only if E and G are lcc and if and only if B and G are lcc.*
(6) *For an lcc sheaf F the function $f : |X| \to \mathbb{Z}; x \mapsto \#F_x$ is locally constant.*

DEFINITION 3.14. A *stratification* $\mathcal{S}$ *of* X is a finite number of non-empty locally closed subtopoi of X, called the *strata*, such that X is the disjoint union of $\mathcal{S}$, and such that the closure of each stratum is the union of strata.

Note that stratifications of X can be identified with stratifications of $|X|$.

DEFINITION 3.15. If F is a sheaf on X, then we call F *constructible* if there exists a stratification $\mathcal{S}$ of X, such that for each $V \in \mathcal{S}$ we have that $F|V$ is lcc. Such a stratification is said to *trivialize* F, or F is said to be $\mathcal{S}$-*constructible*.

LEMMA 3.16. *Let F be sheaf on X. In case X is noetherian, F is constructible if and only if it is so locally.*

LEMMA 3.17. *Let M be a constant sheaf on X, modeled on a finite set. Let N be a subsheaf of M. Then N is constructible.*

PROOF. By noetherian induction it suffices to prove that there is a non-empty open subtopos $U \subset X$ such that $N|U$ is constant. As a sheaf, we have $M \cong X_1 \amalg \ldots \amalg X_n$, where each X_i, for $i = 1, \ldots, n$ is a copy of the final object of X. For each $i = 1, \ldots, n$ let $N_i = N \cap X_i$, which is a subsheaf of X_i and hence isomorphic to an open subtopos $U_i \subset X$. In other words, we have $N \cong U_1 \amalg \ldots \amalg U_n$. Without loss of generality, we may assume that U_1 is non-empty. If $U_1 \cap U_i = \varnothing$ for every $i > 1$, then we may take $U = U_1$. Otherwise, we may assume that $U_1 \cap U_2$ is non-empty. Again, if $U_1 \cap U_2 \cap U_i = \varnothing$ for every $i > 2$, then we may take $U = U_1 \cap U_2$. Continuing in this manner, we find a non-empty open subtopos $U = U_1 \cap \ldots \cap U_k$, for some $k \leq n$, over which N is constant. $\square$

COROLLARY 3.18. *If X is noetherian, every subsheaf of a constructible sheaf is constructible.*

DEFINITION 3.19. A sheaf of A-modules on X is called *lcc* if it is lcc as a sheaf of sets. The category of lcc sheaves of A-modules on X is denoted by $\mathrm{Mod}_{\mathrm{lcc}}(X, A)$.

PROPOSITION 3.20. *The category* $\mathrm{Mod}_{\mathrm{lcc}}(X, A)$ *is a finite closed subcategory (see Definition 1.2) of* $\mathrm{Mod}(X, A)$.

PROOF. The closedness follows easily from Proposition 3.13. For the finiteness, let F be an lcc sheaf on X and let $\ldots \subset F_{i-1} \subset F_i \subset F_{i+1} \subset \ldots$ be a chain of lcc subsheaves of F. To prove that this chain becomes stationary we may assume that $|X|$ is connected. For any i let $h_i : |X| \to \mathbb{Z}$ denote the function $h_i(P) = \#F_{iP}$. Then h_i is continuous and hence constant. This finishes the proof. $\square$

DEFINITION 3.21. A sheaf of A-modules on X is called *constructible*, if it is constructible as a sheaf of sets. The category of constructible sheaves of A-modules on X is denoted by $\mathrm{Mod}_c(X, A)$.

PROPOSITION 3.22. *If X is noetherian, the category* $\mathrm{Mod}_c(X, A)$ *is a noetherian thick subcategory of* $\mathrm{Mod}(X, A)$.

PROOF. Using Proposition 3.20 we easily show that $\mathrm{Mod}_c(X, A)$ is closed under kernels, cokernels and extensions in $\mathrm{Mod}(X, A)$. Then use Corollary 3.18 to conclude. $\square$

Let us denote by $D^+_{\mathrm{lcc}}(X, A)$ the full subcategory of $D(X, A)$ defined by requiring an object M of $D^+_{\mathrm{lcc}}(X, A)$ to satisfy

(1) $h^i M$ is lcc for all $i \in \mathbb{Z}$.
(2) $h^i M = 0$, for all $i \mathbb{L} 0$.

By Proposition 3.20 the subcategory $D^+_{\mathrm{lcc}}(\,\cdot\,, A) \subset D(\,\cdot\,, A)$ is a cd-structure on $|X|$.

DEFINITION 3.23. The cd-structure on $|X|$ thus defined is called the *canonical cd-structure on X, with respect to A*.

NOTE 3.24. A sheaf of A-modules is constructible if and only if it us constructible as an object of the canonical cd-structure (Definition 2.6). Our notion of constructibility for a sheaf of A-modules differs from both that of 1.9.3 in [**2**, Exp. VI] and that of Definition 2.3 in [**2**, Exp. IX], but is more convenient for our purposes.

3. Constructible ℓ-Adic Sheaves

Let X be a noetherian topos.

PROPOSITION 3.25. *Let R be a noetherian ring. Let E be a constant sheaf of R-modules modeled on a finitely generated R-module. Then E is a noetherian object in the category of sheaves of R-modules on X.*

PROOF. First note that every fiber E_P of the sheaf E is canonically isomorphic to the R-module E. So for every subsheaf of R-modules $F \subset E$ we may consider every fiber F_P as a submodule of E. If $P \to Q$ is a specialization arrow of points of X, then for a subsheaf F of E we get an induced homomorphism $F_Q \to F_P$ which commutes with the embeddings of F_Q and F_P into E. Thus we get that $F_Q \subset F_P$ as submodules of E. Note that in particular, the specialization homomorphism $F_Q \to F_P$ does not depend on the given specialization arrow $P \to Q$.

Let now $F_0 \subset F_1 \subset \ldots$ be a chain of subsheaves of R-modules of E. To prove that it is stationary, we may as well assume that X is irreducible. Recall that a topos X is irreducible, if for any two non-empty objects U and V of X their direct product $U \times V$ is non-empty. A topos X with sufficiently many points is irreducible if and only if $|X|$ is irreducible. Using techniques of Section 9 of [**2**, Exp. VI], it is possible to prove that now we may choose $|X|$ in such a way that it contains a point which specializes to every other point of $|X|$. So let P be such a generic point of $|X|$. Since E is noetherian, the chain of submodules $F_{0P}, F_{1P}, \ldots$ is stationary and we may hence assume that $F_{0P} = F_{iP}$ for all i. The R-module F_{0P} is generated by a finite set of elements, say $s_{1P}, \ldots, s_{nP} \in F_{0P}$, where $s_1, \ldots, s_n$ are elements of the R-module E.

Now there exists a non-empty open subtopos $U \subset X$ such that every $s_k : U \to E$ factors through F_0, for example $s_1^{-1}(F_0) \cap \ldots \cap s_n^{-1}(F_0)$. Replacing X by U we may thus assume that F_0 is generated by the global sections $s_1, \ldots, s_n$.

Now let Q be a specialization of P. Then for every $i \geq 0$ we get the following commutative diagram

$$\begin{array}{ccc} F_{iQ} & \subset & F_{iP} \\ \cup & & \| \\ F_{0Q} & = & F_{0P} \end{array}$$

of submodules of E. Note that $F_{0Q} \hookrightarrow F_{0P}$ is an isomorphism, since the sections $s_1, \ldots, s_n$ that generate F_{0P} are global sections of F_0. It follows that $F_{0Q} = F_{iQ}$ for all $Q \in |X|$ and thus that $F_0 = F_i$. □

For the next result note that every locally closed subtopos of a noetherian topos is noetherian (see Proposition 4.6 in [**2**, Exp. VI]).

COROLLARY 3.26. *If F is a locally constant sheaf of R-modules on X, modeled on a finitely generated R-module E, then F is a noetherian object of $\mathrm{Mod}(X, R)$.*

PROOF. Let U be a noetherian object of X such that $F|U$ is constant. The existence of such a U is due to the fact that X is noetherian. By the previous proposition $F|U$ is a noetherian object in $\mathrm{Mod}(U, R)$. This clearly implies that F is a noetherian object of $\mathrm{Mod}(X, R)$. □

LEMMA 3.27. *Let $R = \bigoplus_{p \geq 0} R^p$ be a graded noetherian ring, such that R^0 is finite and for every $p \geq 0$ the R^0-modules R^p and R^0 are isomorphic. Let $M \in \mathrm{ob}\,\mathrm{Mod}(X, R)$ be a graded sheaf of R-modules $M = \bigoplus_{p \geq 0} M^p$ such that each M^p is constructible. Assume that M is a noetherian object of $\mathrm{Mod}(X, R)$. Then there exists a pre-L-stratification $(\mathcal{S}, \mathcal{L})$ of X (with respect to R^0), trivializing M^p, for every $p \geq 0$.*

PROOF. We have a canonical epimorphism $\bigoplus_p R \otimes_{R^0} M^p \to M$ of graded sheaves of R-modules on X. Since M is noetherian, a finite number of the $R \otimes_{R^0} M^p$ will map onto, so there exists an $n \in \mathbb{N}$ such that $\bigoplus_{p=0}^n R \otimes_{R^0} M^p \to M$ is onto.

Note that each homogeneous component of $\bigoplus_{p=0}^n R \otimes_{R^0} M^p$ is constructible. Note also (by Corollary 3.26) that $\bigoplus_{p=0}^n R \otimes_{R^0} M^p$ is a noetherian sheaf of R-modules on X. These two properties clearly carry over to the kernel of $\bigoplus_{p=0}^n R \otimes_{R^0} M^p \to M$. So to this kernel we may apply the same reasoning as for M and obtain constructible sheaves of R^0-modules $N_1, \ldots, N_m$ and an exact sequence of sheaves of R-modules on X

$$\bigoplus_{i=0}^m R \otimes_{R^0} N_i \longrightarrow \bigoplus_{p=0}^n R \otimes_{R^0} M^p \longrightarrow M \longrightarrow 0.$$

Now choose $(\mathcal{S}, \mathcal{L})$ in such a way as to trivialize $N_1, \ldots, N_m$ and $M^0, \ldots, M^n$. Then by Lemma 2.1(ii) of [**2**, Exp. IX] $\mathcal{S}$ trivializes M, and hence M^p for every $p \geq 0$. It is then clear that $(\mathcal{S}, \mathcal{L})$ trivializes M^p for every $p \geq 0$. □

We now turn to the study of constructible ℓ-adic sheaves. So let A be a discrete valuation ring as in Example 3.3. Lift for the moment the assumption that S is noetherian. (Assume only that $|X|$ is noetherian.) We endow the topos X with the canonical A-cd-structure (Definition 3.23). So it makes sense to talk about pre-L-stratifications.

DEFINITION 3.28. We call a sheaf $F = (F_n)_{n \in \mathbb{N}}$ of Λ-modules on $X^{\mathbb{N}}$ *constructible*, if for every $n \in \mathbb{N}$ the component F_n is constructible. The category of constructible sheaves of Λ-modules on $X^{\mathbb{N}}$ will be denoted by $\mathrm{Mod}_c(X^{\mathbb{N}}, \Lambda)$.

Let $(\mathcal{S}, \mathcal{L})$ be a pre-L-stratification of X. A constructible sheaf $F = (F_n)_{n \in \mathbb{N}}$ of Λ-modules on $X^{\mathbb{N}}$ is called $(\mathcal{S}, \mathcal{L})$-*constructible* or *trivialized by* $(\mathcal{S}, \mathcal{L})$, if for every $n \in \mathbb{N}$ the component F_n is trivialized by $(\mathcal{S}, \mathcal{L})$. The category of $(\mathcal{S}, \mathcal{L})$-constructible sheaves of Λ-modules on $X^{\mathbb{N}}$ will be denoted by $\mathrm{Mod}_{(\mathcal{S}, \mathcal{L})}(X^{\mathbb{N}}, \Lambda)$.

The categories of constructible ℓ-adic sheaves and $(\mathcal{S}, \mathcal{L})$-constructible ℓ-adic sheaves are denoted by $\mathrm{Mod}_{\ell, c}(X^{\mathbb{N}}, \Lambda)$ and $\mathrm{Mod}_{\ell, (\mathcal{S}, \mathcal{L})}(X^{\mathbb{N}}, \Lambda)$, respectively. Alternatively, these categories are called the *category of constructible A-sheaves* and the *category of $(\mathcal{S}, \mathcal{L})$-construcible A-sheaves on X*, respectively, and are denoted by $\mathrm{Mod}_c(X, A)$ and $\mathrm{Mod}_{(\mathcal{S}, \mathcal{L})}(X, A)$.

PROPOSITION 3.29. *Let X be a noetherian topos. Let $M = (M_n)_{n \in \mathbb{N}}$ be a constructible ℓ-adic sheaf on X. Then there exists a pre-L-stratification $(\mathcal{S}, \mathcal{L})$ of X such that M is $(\mathcal{S}, \mathcal{L})$-constructible, i.e. $(\mathcal{S}, \mathcal{L})$ simultaneously trivializes M_n, for every $n \in \mathbb{N}$.*

PROOF. For every $n \in \mathbb{N}$ the sheaf of A-modules M_n is a noetherian object of $\mathcal{C} = \mathrm{Mod}(X, A)$, by Proposition 3.22. Thus M is a noetherian ℓ-adic projective system in $\mathcal{C}$, in the sense of Section 5 of [**13**, Exp. V]. So by

Proposition 5.1.6 of [loc. cit.] the strict graded $\mathrm{grs}(M)$ of M is a noetherian object in $\mathrm{Mod}(X, \mathrm{gr}\, A)$, the category of sheaves of $\mathrm{gr}\, A$-modules on X.

Here $\mathrm{gr}\, A = \bigoplus_{p \geq 0} \mathfrak{l}^p/\mathfrak{l}^{p+1}$ is the graded ring associated to the $\mathfrak{l}$-adic filtration of A, $\mathfrak{l}$ being the maximal ideal of A. The strict graded $\mathrm{grs}(M)$ of M is defined as follows: Fix an $n \in \mathbb{N}$. Then consider the $\mathfrak{l}$-adic filtration $F^\bullet M_n$ of M_n given by $F^p M_n = \mathfrak{l}^p M_n$, for $p \geq 0$. By defining $(F^p M)_n = F^p M_n$ we get a filtration $F^\bullet M$ on M. The associated graded is given by

$$(\mathrm{gr}^p M)_n = \mathrm{gr}^p M_n = F^p M_n / F^{p+1} M_n = \mathfrak{l}^p M_n / \mathfrak{l}^{p+1} M_n.$$

Note that if $m \geq n > p$ then the canonical map $\mathfrak{l}^p M_m / \mathfrak{l}^{p+1} M_m \to \mathfrak{l}^p M_n / \mathfrak{l}^{p+1} M_n$ is an isomorphism. Thus $(\mathrm{gr}^p M)_n$ is essentially constant and its limit is defined to be $\mathrm{grs}^p(M)$. So we have $\mathrm{grs}^p M = \mathfrak{l}^p M_n / \mathfrak{l}^{p+1} M_n$ for any $n > p$. Finally, $\mathrm{grs}\, M = \bigoplus_{p \geq 0} \mathrm{grs}^p M$, which is a sheaf of graded $\mathrm{gr}\, A$-modules on X.

By Lemma 3.27 there exists a pre-L-stratification $(\mathcal{S}, \mathcal{L})$ of X that trivializes $\mathrm{grs}^p M$ for every p. Now (for a fixed n) the factors of the $\mathfrak{l}$-adic filtration $F^\bullet M_n$ are $\mathrm{grs}^0 M, \ldots, \mathrm{grs}^{n-1} M$ and so M_n is trivialized by $(\mathcal{S}, \mathcal{L})$. $\square$

COROLLARY 3.30. *The category $\mathrm{Mod}_c(X, A)$ is the union of the full subcategories $\mathrm{Mod}_{(\mathcal{S},\mathcal{L})}(X, A)$, where $(\mathcal{S}, \mathcal{L})$ runs over all pre-L-stratifications of X.*

EXAMPLE 3.31. Let us denote by circ the topos of objects (M, f), where M is a set and f is a permutation of M, such that for every $x \in M$ there exists an $n > 0$ such that $f^n x = x$. Let $i^* : \mathrm{circ} \to pt$ denote the underlying set functor, which forgets the permutation. Then i^* is the fiber functor of a point $i : pt \to \mathrm{circ}$. The set $\{i\}$ is a conservative set of points. Thus there are no non-trivial stratifications of circ. The constructible sheaves are those objects (M, f), for which M is finite. It is a consequence of Proposition 1.13 that there is an equivalence of categories between $\mathrm{Mod}_c(\mathrm{circ}, A)$ and the category of automorphisms of finitely generated $\hat{A}$-modules. For every $n \in \mathbb{Z}$ there exists a morphism of topoi $\epsilon_n : \mathrm{circ} \to \mathrm{circ}$ such that $\epsilon_n^*(M, f) = (M, f^n)$. The induced morphism $\epsilon_n^* : \mathrm{Mod}_c(\mathrm{circ}, A) \to \mathrm{Mod}_c(\mathrm{circ}, A)$ takes an automorphism to its n-th power.

4. Topoi with c-Structures and ℓ-Adic Derived Categories

REMARK 3.32. Since in the applications we have in mind, we do not always have morphisms of topoi available, we introduce the notion of *pseudo-morphism* of topoi. Let X and Y be topoi. Let $f^* : Y \to X$ and $f_* : X \to Y$ be a pair of adjoint functors (f^* being left adjoint of f_*). Then we say that the pair $f = (f^*, f_*)$ is a *pseudo-morphism* from the topos X to the topos Y.

If f is a pseudo-morphism from X to Y, then by definition f^* is continuous. If γ is a species of algebraic structure defined by finite projective limits, then since f_* is left exact, it maps γ-objects to γ-objects. Let us denote by $f_*^\gamma : X_\gamma \to Y_\gamma$ the induced functor on the γ-objects. By Proposition 1.7 of

[**2**, Exp. III] f_*^γ has a left adjoint $f_\gamma^* : Y_\gamma \to X_\gamma$. But in general, f_γ^* does not commute with the underlying-sheaf-of-sets functor.

If (X, A) and (Y, B) are ringed, and $f : X \to Y$ is a pseudo-morphism of ringed topoi, i.e. a pseudo-morphism of topoi together with a ring map $B \to f_*A$, then $f_* : X \to Y$ maps sheaves of A-modules to sheaves of B-modules, so induces a left exact functor $f_* : \text{Mod}(X, A) \to \text{Mod}(Y, B)$. The higher direct image $R^q f_* F$, for a sheaf of A-modules F on X, is the sheaf associated to the presheaf
$$U \longmapsto H^q(f^*U, F),$$
just as in the case of an earnest morphism of topoi. Thus $R^q f_*$ commutes with restriction of scalars.

DEFINITION 3.33. Let X be a topos. A *c-structure* on X is a full subcategory $\overline{X}$ of X such that
(1) $\overline{X}$ is a topos,
(2) the inclusion functor $\pi^* : \overline{X} \to X$ is exact and has a right adjoint $\pi_* : X \to \overline{X}$ such that $\pi_* \pi^* = \text{id}_{\overline{X}}$,
(3) if $P \to B$ is a principal G-bundle in X, then $B \in \text{ob}\,\overline{X}$ and $G \in \text{ob}\,\overline{X}$ implies that $P \in \text{ob}\,\overline{X}$.

To abbreviate, we often call a topos with c-structure a *c-topos*.

A c-topos $(X, \overline{X}, \pi)$ is called *trivial*, if $\pi^* : \overline{X} \to X$, in addition to being continuous, is also co-continuous. We call $(X, \overline{X})$ *quasi-trivial*, if for every abelian sheaf F on $\overline{X}$, π^*F is acyclic for π_*.

A c-structure is called *noetherian*, if the topos $\overline{X}$ is noetherian.

LEMMA 3.34. *Let X be a trivial c-topos. Then π_* has a right adjoint, which we denote by $\pi^!$. Thus X is quasi-trivial.*

PROOF. For the existence of $\pi^!$ see Proposition 2.3 of [**2**, Exp. III]. □

NOTE 3.35. Let X be a topos and $\overline{X}$ a full subcategory such that
(1) $\overline{X}$ contains the final object.
(2) $\overline{X}$ is closed under fibered products in X.
(3) $\overline{X}$ is closed under arbitrary disjoint sums in X.
(4) $\overline{X}$ is closed under quotients of equivalence relations.

Then $\overline{X}$ is a topos, the inclusion functor $\overline{X} \to X$ is exact and has a right adjoint. So if $\overline{X}$ satisfies in addition Condition (3) of Definition 3.33, then it is a c-structure on X.

In particular, we note that any intersection of c-structures is a c-structure. If $\overline{X}$ and $\widetilde{X}$ are two c-structures on X such that $\overline{X} \subset \widetilde{X}$, then $\overline{X}$ is a c-structure on $\widetilde{X}$. If $\overline{X}$ is a c-structure on $\widetilde{X}$ and $\widetilde{X}$ is a c-structure on X, then $\overline{X}$ is a c-structure on X.

EXAMPLE 3.36. Let $X_\bullet$ be a fibered topos. Then $(\text{top}(X_\bullet), X_{\bullet\,\text{cart}})$ is a c-topos. This follows from the fact that any morphism between principal

G-bundles is an isomorphism. The c-topos of Example 3.1 is a quasi-trivial c-topos by Lemma 3.2.

DEFINITION 3.37. Let I be a category. A *fibered c-topos over I* is a fibered topos $X_\bullet$ over I together with a full fibered subcategory $\overline{X}_\bullet$, such that for each $n \in \operatorname{ob} I$ the fiber $\overline{X}_n$ is a c-structure on X_n.

NOTE 3.38. Let $(X_\bullet, \overline{X}_\bullet)$ be a fibered c-topos over I. Then $(\operatorname{top}(X_\bullet), \operatorname{top}(\overline{X}_\bullet))$ is a c-topos. If $(X_n, \overline{X}_n)$ is quasi-trivial for every $n \in \operatorname{ob} I$, then $(\operatorname{top}(X_\bullet), \operatorname{top}(\overline{X}_\bullet))$ is quasi-trivial, too.

PROPOSITION 3.39. *Let $(X, \overline{X})$ be a c-topos. Let Λ be a sheaf of rings on $\overline{X}$, considered also as a sheaf of rings on X, via π^*. The category $\operatorname{Mod}(\overline{X}, \Lambda)$ is closed (see Definition 1.2) in the abelian category $\operatorname{Mod}(X, \Lambda)$.*

PROOF. This follows directly from the definition, noting that a short exact sequence is just a particular kind of principal bundle. □

DEFINITION 3.40. Let $(X, \overline{X})$ be a c-topos. A sheaf F on X is called *constructible*, if it is an object of $\overline{X}$ and is constructible as such (see Definition 3.15). A *stratification* of X is a stratification of $\overline{X}$.

COROLLARY 3.41. *If $(\mathcal{S}, \mathcal{L})$ is a pre-L-stratification of $\overline{X}$ with respect to the ring A, then $\operatorname{Mod}_{(\mathcal{S},\mathcal{L})}(\overline{X}, A)$ is finite and closed in $\operatorname{Mod}(X, A)$.*

PROOF. This follows immediately from Propositions 3.39 and 3.20. □

We denote by $D_{\operatorname{bar}}(X, \Lambda)$ the full subcategory of the derived category $D(X, \Lambda)$ of $\operatorname{Mod}(X, \Lambda)$, defined by requiring the cohomology objects of $M \in \operatorname{ob} D_{\operatorname{bar}}(X, \Lambda)$ to be in $\overline{X}$. By Proposition 1.3 the category $D_{\operatorname{bar}}(X, \Lambda)$ is a t-category with heart $\operatorname{Mod}(\overline{X}, \Lambda)$.

Let A be a ring and assume $\overline{X}$ to be noetherian. Then we denote by $D_c(X, A)$ the subcategory of $D_{\operatorname{bar}}(X, A)$ defined by requiring the cohomology objects $h^i M$ of M to be constructible sheaves on $\overline{X}$, simultaneously trivialized by a common pre-L-stratification of $\overline{X}$. The category $D_c(X, A)$ is an A-t-category with heart $\operatorname{Mod}_c(\overline{X}, A)$. If $(\mathcal{S}, \mathcal{L})$ is a pre-L-stratification of $\overline{X}$, then we denote by $D_{(\mathcal{S},\mathcal{L})}(X, A)$ the subcategory of $D_c(X, A)$ of $(\mathcal{S}, \mathcal{L})$-contructible objects. By definition, we have

$$D_c(X, A) = \bigcup_{(\mathcal{S},\mathcal{L})} D_{(\mathcal{S},\mathcal{L})}(X, A).$$

PROPOSITION 3.42. *Let $(X, \overline{X})$ be a quasi-trivial c-topos and Λ a sheaf of rings on $\overline{X}$. Then the functor $R\pi_* : D_{\operatorname{bar}}^+(X, \Lambda) \to D^+(\overline{X}, \Lambda)$ is an equivalence of categories with quasi-inverse π^*.*

As a consequence, we have for any object $M \in \operatorname{ob} D^+(\overline{X}, \Lambda)$ that $H^i(\overline{X}, M) = H^i(X, \pi^* M)$, for all i.

PROOF. By assumption, the equation $\pi_* \circ \pi^* = \operatorname{id}_{\overline{X}}$ derives to give the equation $R\pi_* \circ \pi^* = \operatorname{id}$ of functors between $D^+(\overline{X}, \Lambda)$ and $D^+(X, \Lambda)$. This proves one direction.

For the other direction, let $M \in \operatorname{ob} D^+_{\operatorname{bar}}(X, \Lambda)$. We need to show that $\pi^* R\pi_* M \to M$ is a quasi-isomorphism. Consider the spectral sequence $R^i \pi_* h^j M \Rightarrow h^{i+j} R\pi_* M$. Since $h^j M$ comes via π^* from $\overline{X}$, by $R\pi_* \circ \pi^* = \operatorname{id}$ this spectral sequence degenerates and we get $\pi_* h^i M = h^i R\pi_* M$, for all i. Thus

$$\begin{aligned} h^i(\pi^* R\pi_* M) &= \pi^* h^i(R\pi_* M) \\ &= \pi^* \pi_* h^i M \\ &= h^i M, \end{aligned}$$

and we are done. □

DEFINITION 3.43. We define three different kinds of morphisms of c-topoi. Let $(X, \overline{X}, \pi_X)$ and $(Y, \overline{Y}, \pi_Y)$ be topoi with c-structure. A *morphism of topoi with c-structures* is a pair $(f, \overline{f})$, where $f : X \to Y$ is a pseudo-morphism of topoi and $\overline{f} : \overline{X} \to \overline{Y}$ is a morphism of topoi such that $\pi_Y \circ f = \overline{f} \circ \pi_X$.

If f is a morphism of topoi (i.e. if f^* is exact), then we call $(f, \overline{f})$ a morphism of c-topoi *of the first kind*.

Now let Λ be a sheaf of rings on $\overline{Y}$. If for every sheaf $F \in \operatorname{ob} \operatorname{Mod}(\overline{X}, \overline{f}^* \Lambda)$ the higher direct images $R^p \overline{f}_* F$ are objects of $\overline{Y}$, for all $p \geq 0$, then we call $(f, \overline{f})$ a morphism of c-topoi *of the second kind* (with respect to Λ).

Finally, let $\overline{X}$ and $\overline{Y}$ be noetherian. Let A be a ring. We call $(f, \overline{f})$ of *the third kind with respect to* A, if the morphism induced by f_*

$$Rf_* : D^+(X, A) \longrightarrow D^+(Y, A)$$

has the following property. Whenever $(\mathcal{S}, \mathcal{L})$ is a pre-L-stratification of $\overline{X}$, then there exists a pre-L-stratification $(\mathcal{S}', \mathcal{L}')$ of $\overline{Y}$ such that if $M \in \operatorname{ob} D^+_{(\mathcal{S}, \mathcal{L})}(X, A)$, then $Rf_* M \in \operatorname{ob} D^+_{(\mathcal{S}', \mathcal{L}')}(Y, A)$.

Now let $f : X \to Y$ be a morphism of c-topoi and A a ring. Let Λ be a sheaf of A-algebras on $\overline{Y}$ and let us denote also by Λ the pullback to X.

If f is of the first kind we get a functor $f^* : D(Y, \Lambda) \to D(X, \Lambda)$ which induces a functor $f^* : D_{\operatorname{bar}}(Y, \Lambda) \to D_{\operatorname{bar}}(X, \Lambda)$ which is, of course, A-linear and t-exact.

If f is of the second kind with respect to Λ, then $Rf_* : D^+(X, \Lambda) \to D^+(Y, \Lambda)$ induces a functor $Rf_* : D^+_{\operatorname{bar}}(X, \Lambda) \to D^+_{\operatorname{bar}}(Y, \Lambda)$, which is A-linear and left t-exact. If X and Y are quasi-trivial with respect to Λ then this Rf_* agrees with $R\overline{f}_*$, via the identifications given by Proposition 3.42. If f is of the first and second kind (f^*, Rf_*) is a pair of adjoint functors between $D^+_{\operatorname{bar}}(X, \Lambda)$ and $D^+_{\operatorname{bar}}(Y, \Lambda)$.

If f is of the third kind with respect to A, we get an induced functor $Rf_* : D^+_c(X, A) \to D^+_c(Y, A)$, which is a right adjoint of f^*, if f is of the first and third kind.

NOTE 3.44. Let $g : Y \to Z$ be another morphism of c-topoi, such that both f and g are of the second kind with respect to Λ. It is not clear whether $g \circ f$ is also of the second kind. Even if this is the case, the question whether $R(g \circ f)_*M \to Rg_*Rf_*M$ is an isomorphism in $D^+_{\text{bar}}(Z, \Lambda)$, for an object M of $D^+_{\text{bar}}(X, \Lambda)$, arises. We do not know if either of these questions can be answered affirmatively without further assumptions. Similar problems arise for morphisms of the third kind.

We will now define the constructible ℓ-adic derived category of a c-topos. Let A be a discrete valuation ring as in Example 3.3. Let $(X, \overline{X})$ be a *noetherian* c-topos. By Corollary 3.41, the following definition is possible.

DEFINITION 3.45. Returning to the notation $\mathfrak{A} = \text{Mod}(X, A)$ and $\mathfrak{A}_c = \text{Mod}_{(\mathcal{S},\mathcal{L})}(\overline{X}, A)$ we get as in Section 2 a t-category $\mathbb{D}_c(\mathfrak{A})$, which we will denote $\mathbb{D}_{(\mathcal{S},\mathcal{L})}(X, A)$ in our case, and call (by abuse of language) the *derived category of $(\mathcal{S}, \mathcal{L})$-constructible A-complexes on the c-topos $(X, \overline{X})$*. By Proposition 1.12, the heart of $\mathbb{D}_{(\mathcal{S},\mathcal{L})}(X, A)$ is $\text{Mod}_{(\mathcal{S},\mathcal{L})}(\overline{X}, A)$, the category of $(\mathcal{S}, \mathcal{L})$-constructible A-sheaves on $\overline{X}$.

Recall that $\mathbb{D}_{(\mathcal{S},\mathcal{L})}(X, A)$ is constructed from $D(X^{\mathbb{N}}, \Lambda)$ by first passing to the subcategory defined by requiring the cohomology to be $(\mathcal{S}, \mathcal{L})$-constructible and AR-$\ell$-adic and then passing to the quotient category modulo those objects whose cohomology is AR-null.

Let $\mathcal{M}$ denote the set of all pre-L-stratifications of $\overline{X}$. The set $\mathcal{M}$ is directed if we call $(\mathcal{S}, \mathcal{L}) \leq (\mathcal{S}', \mathcal{L}')$ if $(\mathcal{S}', \mathcal{L}')$ is a refinement (see Definition 2.9) of $(\mathcal{S}, \mathcal{L})$. In this case we have a natural functor $\mathbb{D}_{(\mathcal{S},\mathcal{L})}(X, A) \to \mathbb{D}_{(\mathcal{S}',\mathcal{L}')}(X, A)$. In general, we cannot expect this functor to be fully faithful, but it is clearly t-exact. On the other hand, the induced functor on the hearts $\text{Mod}_{(\mathcal{S},\mathcal{L})}(\overline{X}, A) \to \text{Mod}_{(\mathcal{S}',\mathcal{L}')}(\overline{X}, A)$ is obviously fully faithful.

DEFINITION 3.46. We define the *category of constructible A-complexes on the c-topos X* to be the 2-limit

$$\mathbb{D}_c(X, A) = \underset{(\mathcal{S},\mathcal{L})\in\mathcal{M}}{\text{inj lim}} \mathbb{D}_{(\mathcal{S},\mathcal{L})}(X, A).$$

PROPOSITION 3.47. *The A-category $\mathbb{D}_c(X, A)$ is naturally a t-category with heart $\text{Mod}_c(\overline{X}, A)$.*

PROOF. This follows easily from the definitions and Corollary 3.30. □

REMARK 3.48. Let $(X, \overline{X})$ and $(Y, \overline{Y})$ be noetherian c-topoi and $f : X \to Y$ a morphism of c-topoi of the first kind. Let $(\mathcal{S}, \mathcal{L})$ be a pre-L-stratification of $\overline{Y}$. Let $\mathcal{S}'$ be the pullback of $\mathcal{S}$ to $\overline{X}$. For a stratum $\overline{W} \in \mathcal{S}$ let $\mathcal{L}(\overline{W}) = \{L_1, \ldots, L_n\}$. Pulling back $L_1, \ldots, L_n$ to $\overline{V} = f^*\overline{W}$ gives a collection $f^*L_1, \ldots, f^*L_n$ of lcc sheaves of A-modules on $\overline{V}$, which we may decompose into Jordan-Hölder components, arriving at a collection of simple lcc sheaves on $\overline{V}$, which we call $\mathcal{L}'(\overline{V})$. Then $(\mathcal{S}', \mathcal{L}')$ is a pre-L-stratification of $\overline{X}$, which we shall call the pullback of the pre-L-stratification $(\mathcal{S}, \mathcal{L})$ of $\overline{Y}$. Clearly, the functor

$$f^* : D_{\mathrm{bar}}(Y^{\mathbb{N}}, \Lambda) \longrightarrow D_{\mathrm{bar}}(X^{\mathbb{N}}, \Lambda)$$

induces a functor

$$f^* : D_{(\mathcal{S},\mathcal{L})}(Y^{\mathbb{N}}, \Lambda) \longrightarrow D_{(\mathcal{S}',\mathcal{L}')}(X^{\mathbb{N}}, \Lambda),$$

which induces a functor

$$f^* : \mathbb{D}_{(\mathcal{S},\mathcal{L})}(Y, A) \longrightarrow \mathbb{D}_{(\mathcal{S}',\mathcal{L}')}(X, A).$$

Passing to the limit we get an induced A-linear t-exact functor

$$f^* : \mathbb{D}_c(Y, A) \longrightarrow \mathbb{D}_c(X, A).$$

REMARK 3.49. Let $(X, \overline{X})$ and $(Y, \overline{Y})$ be noetherian c-topoi and $f : X \to Y$ a morphism of c-topoi of the third kind, with respect to A. Let $(\mathcal{S}, \mathcal{L})$ be a pre-L-stratification of $\overline{X}$ and $(\mathcal{S}', \mathcal{L}')$ a pre-L-stratification of $\overline{Y}$ satisfying Definition 3.43. By Corollary 1.17 the derived functor

$$Rf_*^{\mathbb{N}} : D_{(\mathcal{S},\mathcal{L})}^+(X^{\mathbb{N}}, \Lambda) \longrightarrow \mathbb{D}_{(\mathcal{S}',\mathcal{L}')}^+(Y^{\mathbb{N}}, \Lambda)$$

induces a left t-exact functor

$$\mathbb{R}f_* : \mathbb{D}_{(\mathcal{S},\mathcal{L})}^+(X, A) \longrightarrow \mathbb{D}_{(\mathcal{S}',\mathcal{L}')}^+(Y, A).$$

Passing to the limit we get an induced left t-exact A-linear functor

$$\mathbb{R}f_* : \mathbb{D}_c^+(X, A) \longrightarrow \mathbb{D}_c^+(Y, A).$$

We call $\mathbb{R}f_*$ the *ℓ-adic derived functor of* $f_* : \mathrm{Mod}_c(X, A) \to \mathrm{Mod}_c(Y, A)$.

EXAMPLE 3.50. Let $X = pt$ be the punctual topos and assume that A has finite residue field. Then $\mathrm{Mod}(X, A)$ is just the category of A-modules and $\mathrm{Mod}_c(X, A)$ is the category of finite A-modules. Every constructible sheaf of A-modules is lcc. Up to isomorphism, the only simple finite A-module is the residue field of A. So there is only one pre-L-stratification of X and every constructible A-module is trivialized by it. Thus $\mathbb{D}_c(pt, A) = D_{\mathrm{fg}}(\hat{A})$, by Proposition 1.14.

5. The d-Structures Defined by c-Topoi

We will now study the way in which c-topoi give rise to d-structures. Let $(X, \overline{X})$ be a c-topos with structure morphism $\pi : X \to \overline{X}$. Let $\overline{V}$ be a locally closed subtopos of $\overline{X}$ and $V = \pi^{-1}(\overline{V})$. Denote by ρ the induced morphism $\rho : V \to \overline{V}$.

PROPOSITION 3.51. *The functor ρ^* is fully faithful. The essential image of ρ^* is a c-structure on V.*

5. THE D-STRUCTURES DEFINED BY C-TOPOI

PROOF. We may assume that $\overline{V}$ is closed or open in $\overline{X}$. Denote by X_0 a final object of X and by U_0 a subobject of X_0 which is an object of $\overline{X}$. Then U_0 gives rise to an open subtopos $\overline{U}$ of $\overline{X}$ with closed complement $\overline{Z}$. Let $U = \pi^{-1}(\overline{U})$ and $Z = \pi^{-1}(\overline{Z})$. We have a 2-cartesian diagram of topoi

$$\begin{array}{ccc} X_{/U_0} & \xrightarrow{j} & X \\ \rho \downarrow & & \downarrow \pi \\ \overline{X}_{/U_0} & \xrightarrow{\overline{j}} & \overline{X}, \end{array}$$

where ρ^* is the inclusion functor of the subcategory $\overline{X}_{/U_0}$ of $X_{/U_0}$ (U and $\overline{U}$ are the essential images of j_* and $\overline{j}_*$, respectively). So we need to prove that $\overline{X}_{/U_0}$ is a c-structure on $X_{/U_0}$. But this is clear, after a moments thought.

An object F of X is in Z if and only if $F \times U_0 \to U_0$ is an isomorphism. It is in $\overline{Z}$ if and only if it is in Z and in $\overline{X}$. The fact that $\overline{Z}$ is a c-structure on Z follows from the fact that the exact functor $i^* : X \to Z$ maps $\overline{X}$ into $\overline{Z}$. □

DEFINITION 3.52. We call $(V, \overline{V})$ the *induced c-topos over the locally closed subtopos* $\overline{V}$ *of* $\overline{X}$.

WARNING 3.53. The diagram of 'inclusion' functors

$$\begin{array}{ccc} V & \xrightarrow{j_*} & X \\ \rho^* \uparrow & & \uparrow \pi^* \\ \overline{V} & \xrightarrow{\overline{j}_*} & \overline{X} \end{array} \qquad (4)$$

does not necessarily commute.

NOTE 3.54. The pair $(j, \overline{j}) : (V, \overline{V}) \to (X, \overline{X})$ is a morphism of c-topoi of the first kind.

Let X be a topos. Choosing for every pair $V \subset W$ of locally closed subtopoi of X a pullback functor $i^* : W \to V$ defines a fibered topos over $\mathrm{lc}(X)$, the ordered set of locally closed subtopoi of X, whose fiber over $V \in \mathrm{lc}(X)$ if V itself.

Now let $\overline{X}$ be a c-structure on X. The morphism $\pi : X \to \overline{X}$ defines a map $\pi^{-1} : \mathrm{lc}(\overline{X}) \to \mathrm{lc}(X)$ which preserves inclusions and may thus be considered as a functor. Pulling our fibered topos $V \mapsto V$ back via π^{-1} defines a fibered topos over $\mathrm{lc}(\overline{X})$, which we will denote by $\overline{V} \mapsto V$. Endow every fiber of $\overline{V} \mapsto V$ with its induced c-structure according to Definition 3.52. This procedure gives rise to a fibered c-topos over $\mathrm{lc}(\overline{X})$, denoted by $\overline{V} \mapsto (V, \overline{V})$.

Now let Λ be a sheaf of A-algebras on $\overline{X}$. To every object $\overline{V} \in \mathrm{lc}(\overline{X})$ assign the A-t-category $D^+_{\mathrm{bar}}(V, \Lambda)$, associated to the c-topos $(V, \overline{V})$. To every inclusion $\overline{j} : \overline{V} \to \overline{W}$ in $\mathrm{lc}(\overline{X})$ assign the pullback functor $j^* : D^+_{\mathrm{bar}}(W, \Lambda) \to D^+_{\mathrm{bar}}(V, \Lambda)$, associated to the morphism of c-topoi of the first kind $(j, \overline{j}) : (V, \overline{V}) \to (W, \overline{W})$. This construction defines a category $D^+_{\mathrm{bar}}(\,\cdot\,, \Lambda)$, fibered in A-t-categories over $\mathrm{lc}(\overline{X})$.

Now let X_0 be a noetherian topological space (for example $X_0 = |\overline{X}|$, if $\overline{X}$ is noetherian) and $\gamma^{-1} : \operatorname{lc} X_0 \to \operatorname{lc} \overline{X}$ a functor preserving open and closed immersions and taking $\varnothing$ to $\varnothing$. For example, γ^{-1} may be induced by a continuous map $\gamma : |\overline{X}| \to X_0$, if $\overline{X}$ has sufficiently many points. Then via γ^{-1}, we may pull back $D^+_{\mathrm{bar}}(\,\cdot\,, \Lambda)$ to get a fibered category over $\operatorname{lc} X_0$.

PROPOSITION 3.55. *Assume that for every open immersion $V_0 \to W_0$ in $\operatorname{lc}(X_0)$ the associated morphism of c-topoi $(V, \overline{V}) \to (W, \overline{W})$ is of the second kind with respect to Λ. Then the above procedure gives rise to a d-structure $D^+_{\mathrm{bar}}(\,\cdot\,, \Lambda)$ on X_0.*

PROOF. We have the d-structure $D^+(\,\cdot\,, \Lambda)$ and so we only have to check that $D^+_{\mathrm{bar}}(\,\cdot\,, \Lambda)$ is preserved by k^*, k_*, $k^!$, and $k_!$, for $k : V_0 \to W_0$ a morphism in $\operatorname{lc} X_0$. But this is easily done. □

EXAMPLE 3.56. We may apply these considerations to the c-topos $(X^{\mathbb{N}}, \overline{X}^{\mathbb{N}})$. We let A be a discrete valuation ring and $\Lambda = (A/\ell^{n+1})_{n \in \mathbb{N}}$. We will assume that $\overline{X}$ is noetherian and take $X_0 = |\overline{X}|$. The map $\gamma^{-1} : \operatorname{lc}|\overline{X}| \to \operatorname{lc}\overline{X}^{\mathbb{N}}$ is just the one induced by the morphism of topoi $\overline{X}^{\mathbb{N}} \to \overline{X}$. Finally, let us assume that for every open immersion $\overline{V} \subset \overline{W}$ in $\operatorname{lc}(\overline{X})$ the induced morphism of c-topoi $(V, \overline{V}) \to (W, \overline{W})$ is of the second kind with respect to A/ℓ^{n+1}, for every $n \in \mathbb{N}$. This assumption implies that $(V^{\mathbb{N}}, \overline{V}^{\mathbb{N}}) \to (W^{\mathbb{N}}, \overline{W}^{\mathbb{N}})$ is of the second kind with respect to Λ. So by Proposition 3.55 we get a d-structure on $|\overline{X}|$, whose fiber over $\overline{V} \in \operatorname{lc}(\overline{X})$ is $D^+_{\mathrm{bar}}(V^{\mathbb{N}}, \Lambda)$.

PROPOSITION 3.57. *Let $\overline{X}$ be noetherian and assume that for all open immersions $\overline{V} \subset \overline{W}$ in $\operatorname{lc}(\overline{X})$ the corresponding morphism of c-topoi $(V, \overline{V}) \to (W, \overline{W})$ is of the third kind with respect to A. Then $\overline{V} \mapsto D^+_c(V, A)$ defines a d-structure on $|\overline{X}|$. The subcategories $D^+_{\mathrm{lcc}}(V, A) \subset D^+_c(V, A)$ define a tractable cd-structure.*

PROOF. This is proved just like Proposition 3.55. Use Proposition 3.20. □

NOTE 3.58. In this proposition it suffices to assume that for every open immersion $\overline{j} : \overline{V} \to \overline{W}$ in $\operatorname{lc}(\overline{X})$ the corresponding functor $Rj_* : D^+(V, A) \to D^+(W, A)$ maps $D^+_c(V, A)$ into $D^+_c(W, A)$.

PROPOSITION 3.59. *Assume we are in the situation of Example 3.56, with the additional hypothesis of Proposition 3.57 verified. Then the d-structure $\overline{V} \mapsto D^+_{\mathrm{bar}}(V^{\mathbb{N}}, \Lambda)$ induces a d-structure on $|\overline{X}|$*

$$\overline{V} \longmapsto \mathbb{D}^+_c(V, A).$$

PROOF. First let us introduce the following notation. If Y is a noetherian topological space and $\mathcal{S}$ a stratification of Y, then by $Y_{\mathcal{S}}$ we denote the topology on Y in which a subset of Y is open if it is open in Y and the union of strata from $\mathcal{S}$.

Fix a stratification $\mathcal{S}_0$ of $|\overline{X}|$ and let $(\mathcal{S}, \mathcal{L})$ be an L-stratification of $\overline{X}$ such that $\mathcal{S}$ refines $\mathcal{S}_0$. Then $(\mathcal{S}, \mathcal{L})$ induces an L-stratification on every locally closed subtopos $\overline{V}$ of $\overline{X}$, which is the union of strata in $\mathcal{S}_0$ (see Lemma 2.11).

We get an induced d-structure on $|\overline{X}|_{\mathcal{S}_0}$ with fiber $D^+_{(\mathcal{S},\mathcal{L})}(V^{\mathbb{N}}, \Lambda)$, which passes to a d-structure with fiber $\mathbb{D}^+_{(\mathcal{S},\mathcal{L})}(V, A)$ using Remark 3.49 and Proposition 3.8. We will call this d-structure by abuse of notation $\mathbb{D}^+_{(\mathcal{S},\mathcal{L})}(|\overline{X}|_{\mathcal{S}_0}, A)$.

If $(\mathcal{S}', \mathcal{L}')$ is a refinement of $(\mathcal{S}, \mathcal{L})$, then we get a cartesian functor between d-structures on $|\overline{X}|_{\mathcal{S}_0}$

$$\mathbb{D}^+_{(\mathcal{S},\mathcal{L})}(|\overline{X}|_{\mathcal{S}_0}, A) \longrightarrow \mathbb{D}^+_{(\mathcal{S}',\mathcal{L}')}(|\overline{X}|_{\mathcal{S}_0}, A).$$

Taking the limit defines a d-structure $\mathbb{D}^+_c(|\overline{X}|_{\mathcal{S}_0}, A)$ on $|\overline{X}|_{\mathcal{S}_0}$.

Now let $\mathcal{S}_1$ be a refinement of $\mathcal{S}_0$. Then $\mathbb{D}^+_c(|\overline{X}|_{\mathcal{S}_0}, A)$ is just the restriction of $\mathbb{D}^+_c(|\overline{X}|_{\mathcal{S}_1}, A)$ to $\mathrm{lc}\,|\overline{X}|_{\mathcal{S}_0}$. So taking the limit over all stratifications of $|\overline{X}|$ we get the desired d-structure on $|\overline{X}|$. $\square$

6. Topoi with e-Structures

DEFINITION 3.60. Let X be a topos. An *e-structure on X* is a subset E of $\mathrm{fl}(X)$ satisfying

(1) All identities are in E.
(2) If

$$\begin{array}{ccc} F' & \xrightarrow{f'} & G' \\ \downarrow & & \downarrow u \\ F & \xrightarrow{f} & G \end{array}$$

is a cartesian diagram in X, then $f \in E$ implies $f' \in E$. If u is an epimorphism, the converse is true.

(3) If

$$\begin{array}{ccc} F & \xrightarrow{f} & G \\ {}_h\searrow & & \downarrow g \\ & H & \end{array}$$

is a commutative diagram in X, then $f, g \in E$ implies $h \in E$, $g, h \in E$ implies $f \in E$ and if f is an epimorphism then $f, h \in E$ implies $g \in E$.

(4) If $U_i \to V$ is a family of E-morphisms, then so is $\coprod U_i \to V$.

Let X be a topos with an e-structure E. For any object U of X, let $\overline{U}$ be the category of U-objects whose structure morphism is in E.

LEMMA 3.61. *For any $U \in \mathrm{ob}\,X$ the pair $(U, \overline{U})$ is a c-topos. Any morphism $U \to V$ in X gives rise to a morphism of c-topoi of the first kind $(U, \overline{U}) \to (V, \overline{V})$.*

EXAMPLE 3.62. Let X be a topos with e-structure E. Call a morphism $\phi^\bullet : F^\bullet \to G^\bullet$ in $X^\mathbb{N}$ an E-morphism if for every $n \in \mathbb{N}$ the morphism $\phi^n : F^n \to G^n$ is an E-morphism of X. Then the collection of E-morphisms is an e-structure on $X^\mathbb{N}$.

Let X be a topos with e-structure E and let us denote by $\mathcal{U}$ a set of objects of X satisfying

(1) Every $U \in \mathcal{U}$ covers the final object of X.
(2) For every $U \in \mathcal{U}$ the c-topos $(U, \overline{U})$ is quasi-trivial.
(3) $\mathcal{U}$ is closed under finite direct products.

Fix $U \in \mathcal{U}$. Let $U_\bullet$ be the Čech nerve of the covering $U \to X$ (see Example 3.1). Then $(U_\bullet, \overline{U}_\bullet)$ is a fibered c-topos such that $U_{\bullet\mathrm{cart}} = X$ and $\overline{U}_{\bullet\mathrm{cart}} = \overline{X}$. The total topos $\mathrm{top}(U_\bullet)$ has the c-structures $\mathrm{top}(\overline{U}_\bullet)$ and $U_{\bullet\mathrm{cart}} = X$. Both of these are quasi-trivial. The first one by Note 3.38 and the second one by Lemma 3.2. Their intersection $\mathrm{top}(\overline{U}_\bullet) \cap U_{\bullet\mathrm{cart}} = \overline{X}$ is a third c-structure on $\mathrm{top}(U_\bullet)$. Let us denote the structure morphisms by $\pi : \mathrm{top}(U_\bullet) \to X$ and $\sigma : \mathrm{top}(U_\bullet) \to \mathrm{top}(\overline{U}_\bullet)$.

$$\begin{array}{ccc} \mathrm{top}(U_\bullet) & \xrightarrow{\sigma} & \mathrm{top}(\overline{U}_\bullet) \\ \pi \downarrow & & \downarrow \\ X & \longrightarrow & \overline{X} \end{array}$$

Note that both $\sigma_* \pi^* : X \to \mathrm{top}(\overline{U}_\bullet)$ and $\pi_* \sigma^* : \mathrm{top}(\overline{U}_\bullet) \to X$ induce the identity on the c-structure $\overline{X}$.

Now let Λ be a sheaf of rings on $\overline{X}$. Let $D_e(X, \Lambda)$ denote the full subcategory of $D(X, \Lambda)$ given by the c-structure $\overline{X}$ and let $D_{\mathrm{cart}}(\mathrm{top}(\overline{U}_\bullet), \Lambda)$ be the subcategory of $D(\mathrm{top}(\overline{U}_\bullet), \Lambda)$ given by the c-structure $\overline{U}_{\bullet\mathrm{cart}}$.

PROPOSITION 3.63. *The derived functor of $\sigma_* \pi^*$ induces an equivalence of t-categories*
$$R(\sigma_* \pi^*) : D_e^+(X, \Lambda) \longrightarrow D_{\mathrm{cart}}^+(\overline{U}_\bullet, \Lambda).$$

PROOF. Let $D_e(U_\bullet, \Lambda)$, $D_{\mathrm{cart}}(U_\bullet, \Lambda)$ and $D_{e,\mathrm{cart}}(U_\bullet, \Lambda)$ denote the subcategories of $D(U_\bullet, \Lambda)$ given by the c-structures $\mathrm{top}\,\overline{U}_\bullet$, X and $\overline{X}$ on $\mathrm{top}\,U_\bullet$, respectively. By Proposition 3.42 we have equivalences of categories

$$D_e^+(U_\bullet, \Lambda) \underset{\sigma^*}{\overset{R\sigma_*}{\rightleftarrows}} D^+(\overline{U}_\bullet, \Lambda)$$

and

$$D_{\mathrm{cart}}^+(U_\bullet, \Lambda) \underset{\pi^*}{\overset{R\pi_*}{\rightleftarrows}} D^+(X, \Lambda).$$

These trivially induce equivalences

$$D_{e,\mathrm{cart}}^+(U_\bullet, \Lambda) \underset{\sigma^*}{\overset{R\sigma_*}{\rightleftarrows}} D_{\mathrm{cart}}^+(\overline{U}_\bullet, \Lambda)$$

and
$$D^+_{e,\text{cart}}(U_\bullet, \Lambda) \underset{\pi^*}{\overset{R\pi_*}{\rightleftarrows}} D^+_e(X, \Lambda).$$
Composing these, we get the required equivalence
$$D^+_e(X, \Lambda) \underset{R\pi_*\circ\sigma^*}{\overset{R\sigma_*\circ\pi^*}{\rightleftarrows}} D^+_{\text{cart}}(\overline{U}_\bullet, \Lambda).$$

To prove that $R\sigma_* \circ \pi^*$ is induced by the derived functor of $\sigma_*\pi^*$, we have to prove that if F is an injective sheaf of Λ-modules on X, then π^*F is acyclic for σ_*. But this follows from the fact that $R\sigma_*$ may be computed componentwise and every component of π^*F is injective.

Note that $R\pi_*\circ\sigma^*$ is not induced by the derived functor of $\pi_*\sigma^*$, since an injective sheaf of Λ-modules on top $\overline{U}_\bullet$ is not necessarily acyclic for π_*. $\square$

COROLLARY 3.64. *We get an induced equivalence of categories*
$$R(\sigma_*\pi^*) : D^+_e(X^\mathbb{N}, \Lambda) \longrightarrow D^+_{\text{cart}}(\overline{U}_\bullet^\mathbb{N}, \Lambda),$$
for any sheaf of rings Λ on $\overline{X}^\mathbb{N}$. If Λ is as in Example 3.3 and $\overline{X}$ is noetherian, this equivalence further passes to an equivalence of categories
$$R(\sigma_*\pi^*) : \mathbb{D}^+_c(X, A) \longrightarrow \mathbb{D}^+_c(\overline{U}_\bullet, A).$$

PROOF. Apply the proposition to Example 3.62. $\square$

Now let $f : U \to V$ be a morphism in X, where both U and V are elements of $\mathcal{U}$. Then f gives rise to a morphism of c-topoi $(\text{top}\,\overline{U}_\bullet, \overline{X}) \to (\text{top}\,\overline{V}_\bullet, \overline{X})$ of the first kind. Hence we have an induced morphism $f^* : D^+_{\text{cart}}(\overline{V}_\bullet, \Lambda) \to D^+_{\text{cart}}(\overline{U}_\bullet, \Lambda)$. So considering $\mathcal{U}$ as a full subcategory of X, we get a fibered category
$$U \longmapsto D^+_{\text{cart}}(\overline{U}_\bullet, \Lambda)$$
over $\mathcal{U}$.

PROPOSITION 3.65. *There is an equivalence $R(\sigma_*\pi^*)$ of fibered categories over $\mathcal{U}$, from $D^+_e(X, \Lambda)$ to $U \mapsto D^+_{\text{cart}}(\overline{U}_\bullet, \Lambda)$. Here we think of $D^+_e(X, \Lambda)$ as a constant fibered category of $\mathcal{U}$.*

PROOF. Let $U, V \in \mathcal{U}$ and let $f : U \to V$ be a morphism. We need to show that the diagram
$$\begin{array}{ccc}
D^+_e(X, \Lambda) & \overset{R(\sigma_*\pi^*)}{\longrightarrow} & D^+_{\text{cart}}(\overline{U}_\bullet, \Lambda) \\
{\scriptstyle R(\sigma_*\pi^*)} \searrow & & \uparrow f^* \\
& D^+_{\text{cart}}(\overline{V}_\bullet, \Lambda) &
\end{array}$$
naturally commutes. But there is a natural morphism $\theta : f^*\circ\sigma_*\circ\pi^* \to \sigma_*\circ\pi^*$. We wish to see that for $M \in \text{ob}\,D^+_e(X, \Lambda)$ we get a quasi-isomorphism $\theta : f^*R\sigma_*\pi^*M \to R\sigma_*\pi^*M$. This is easily reduced to the case $M \in \text{ob}\,\text{Mod}(\overline{X}, \Lambda)$, for which it is clear. $\square$

CHAPTER 4

Algebraic Stacks

1. Preliminaries on Algebraic Stacks

Throughout this section S will stand for a noetherian scheme, which we will use as a base for our constructions.

1.1. Gerbe-Like morphisms. All relative algebraic group schemes (or spaces) will be assumed to be of finite type. We will sometimes tacitly assume that a relative algebraic group space is a group scheme. This is justified by the following Proposition.

PROPOSITION 4.1. *Let $G \to S$ be a relative algebraic space of groups. Then there exists a stratification $\mathcal{S}$ of S such that $G_V \to V$ is a group scheme for every $V \in \mathcal{S}$.*

PROOF. Without loss of generality G is flat over X and there exists an open subscheme of G mapping onto S. Then from the homogeneous nature of G it is easily seen that G is a scheme, using faithfully flat descent. □

PROPOSITION 4.2. *Let G be a flat group space of finite presentation over the scheme S. Then BG is smooth over S.*

PROOF. It is a result of Artin (see [**1**, Theorem 6.1]), that BG is an algebraic stack. So it has a smooth presentation $p : X \to BG$. The morphism p is induced by a principal G-bundle P over X. The algebraic S-space P is faithfully flat of finite presentation over X and smooth over S, as is obvious from the cartesian diagram

$$\begin{array}{ccc} P & \longrightarrow & S \\ \downarrow & & \downarrow \\ X & \stackrel{p}{\longrightarrow} & BG. \end{array}$$

But smoothness is local with respect to the fppf-topology (see [**15**]), so X is smooth over S. □

DEFINITION 4.3. Let $f : \mathfrak{X} \to \mathfrak{Y}$ be a morphism of algebraic S-stacks. We call f *gerbe-like*, if f and $\Delta_f : \mathfrak{X} \to \mathfrak{X} \times_{\mathfrak{Y}} \mathfrak{X}$ are flat epimorphisms of finite presentation. Let f be gerbe-like. We call f *diagonally connected*, if Δ_f has connected geometric fibers.

NOTE 4.4. The property of being gerbe-like (respectively gerbe-like and diagonally connected) is local on the base with respect to the fppf-topology.

PROPOSITION 4.5. *Every gerbe-like morphism is smooth.*

PROOF. Since the properties of being gerbe-like or smooth are both local with respect to the fppf-topology on $\mathfrak{Y}$ we may assume that $\mathfrak{Y} = Y$ is a scheme and that our morphism has a section. Then $\mathfrak{X}$ is a neutral gerbe over Y, hence isomorphic to $B(G/Y)$, for a flat group space of finite presentation G over Y. Now we use Proposition 4.2. □

DEFINITION 4.6. Let $f : \mathfrak{X} \to \mathfrak{Y}$ be a flat representable morphism of finite type between algebraic S-stacks. We say that f *has components* (or that $\mathfrak{X}$ *has components over* $\mathfrak{Y}$), if f factors as $\mathfrak{X} \to \mathfrak{Y}' \to \mathfrak{Y}$, where $\mathfrak{Y}' \to \mathfrak{Y}$ is representable étale and $\mathfrak{X} \to \mathfrak{Y}'$ is surjective with geometrically connected fibers.

NOTE 4.7. If $f : \mathfrak{X} \to \mathfrak{Y}$ has components, the S-stack $\mathfrak{Y}'$ in this definition is uniquely determined by f. The morphism $\mathfrak{X} \to \mathfrak{Y}'$ is a flat representable epimorphism. The property of having components is stable under base change and local on the base with respect to the fppf-topology.

EXAMPLE 4.8. Let G be and S-group space of finite type with connected fibers. Let X be a flat S-space of finite type on which G acts in such a way that the graph of the action $G \times X \to X \times X$ is an open immersion. Then X has components (over S). In fact, the factorization $X \to X/G \to S$ satisfies the requirements of Definition 4.6.

COROLLARY 4.9. *Let G be a flat S-group space of finite type. Then G has components (over S) if and only if G° is representable. Here G° is the subsheaf of G defined by*

$$G^\circ(T) = \{g \in G(T) \mid g(t) \in G^\circ_t, \text{ for all } t \in T\},$$

for every S-scheme T (see Definition 3.1 in [**10**, *Exp. VIB*]*).*

EXAMPLE 4.10. A smooth S-group scheme G of finite type has components. In fact, according to Théorème 3.10 in [**10**, Exp. VIB], G° is representable if G is a smooth group scheme.

PROPOSITION 4.11. *Let $f : \mathfrak{X} \to \mathfrak{Y}$ be a gerbe-like morphism of algebraic S-stacks such that Δ_f has components. Then f factors in a unique fashion as $\mathfrak{X} \xrightarrow{g} \overline{\mathfrak{X}}_\mathfrak{Y} \xrightarrow{h} \mathfrak{Y}$, where g and h are gerbe-like, h is étale and g is diagonally connected.*

PROOF. Without loss of generality $\mathfrak{Y} = Y$ is a scheme and f has a section. Then there exists a flat group space of finite presentation G over Y such that $\mathfrak{X} \cong B(G/Y)$. Then f factors as $BG \to B(G/G^\circ) \to Y$. Now G/G° is étale over Y, so $B(G/G^\circ)$ is so, too. Moreover, we have a 2-cartesian diagram

$$\begin{array}{ccc} BG^\circ & \longrightarrow & Y \\ \downarrow & & \downarrow \\ BG & \longrightarrow & B(G/G^\circ) \end{array}$$

and $BG^\circ \to Y$ is diagonally connected, so the same is true for $BG \to B(G/G^\circ)$.

Conversely, assume that $BG \to Y$ factors as $BG \to \mathfrak{Z} \to Y$. Then $\mathfrak{Z} \to Y$ has a section, given by the image of the section of $BG \to Y$. Via this section, $\mathfrak{Y} \cong B(G'/Y)$, for some étale group space of finite presentation G' over Y. Now the morphism $BG \to BG'$ induces a morphism on the automorphism groups of the canonical objects of $BG(Y)$ and $BG'(Y)$, respectively, in other words we get an induced homomorphism $G \to G'$, giving rise to the given morphism $BG \to BG'$. Consider the 2-cartesian diagram

$$\begin{array}{ccc} G & \longrightarrow & G' \\ \downarrow & & \downarrow \\ BG & \stackrel{\Delta}{\longrightarrow} & BG \times_{BG'} BG. \end{array}$$

It shows that $BG \to BG'$ being gerbe-like and diagonally connected implies that $G \to G'$ is an epimorphism with connected kernel. Then we have necessarily $G' \cong G/G^\circ$. □

REMARK 4.12. Let $f : \mathfrak{X} \to X$ be gerbe-like such that Δ_f has components and where X is an algebraic S-space. Then X is the coarse moduli space of $\mathfrak{X}$. Let $\mathfrak{X} \to \overline{\mathfrak{X}} \to X$ be the factorization given by the proposition. Then we call $\overline{\mathfrak{X}}$ the *Deligne-Mumford stack associated to* $\mathfrak{X}$.

1.2. Devissage for Algebraic Stacks.

REMARK 4.13. For a morphism $f : \mathfrak{X} \to \mathfrak{Y}$ of algebraic S-stacks we define $\mathfrak{Aut}_\mathfrak{Y} \mathfrak{X}$ by the 2-cartesian diagram

$$\begin{array}{ccc} \mathfrak{Aut}_\mathfrak{Y} \mathfrak{X} & \longrightarrow & \mathfrak{X} \\ \downarrow & & \downarrow \Delta \\ \mathfrak{X} & \stackrel{\Delta}{\longrightarrow} & \mathfrak{X} \times_\mathfrak{Y} \mathfrak{X}. \end{array}$$

If $Y \to \mathfrak{Y}$ is a base change, then $\mathfrak{Aut}_\mathfrak{Y} \mathfrak{X} \times_\mathfrak{Y} Y = \mathfrak{Aut}_Y(\mathfrak{X} \times_\mathfrak{Y} Y)$.

Now assume that $\mathfrak{Aut}_\mathfrak{Y} \mathfrak{X} \to \mathfrak{X}$ is flat. Then $f : \mathfrak{X} \to \mathfrak{Y}$ factors uniquely as $\mathfrak{X} \to \widetilde{\mathfrak{X}}_\mathfrak{Y} \to \mathfrak{Y}$, where $\mathfrak{X} \to \widetilde{\mathfrak{X}}_\mathfrak{Y}$ is gerbe-like and $\widetilde{\mathfrak{X}}_\mathfrak{Y} \to \mathfrak{Y}$ is representable. This is easily seen by noting that for a presentation $Y \to \mathfrak{Y}$ of $\mathfrak{Y}$, the space $\widetilde{\mathfrak{X}}_\mathfrak{Y} \times_\mathfrak{Y} Y$ is necessarily the coarse moduli space of $\mathfrak{X} \times_\mathfrak{Y} Y$.

PROPOSITION 4.14. *Let $f : \mathfrak{X} \to \mathfrak{Y}$ be a finite type morphism of algebraic S-stacks. If $\mathfrak{X}$ is non-empty, there exists a non-empty open substack $\mathfrak{X}'$ of $\mathfrak{X}$ such that $\mathfrak{Aut}_\mathfrak{Y} \mathfrak{X}'$ is flat and has components over $\mathfrak{X}'$.*

PROOF. Without loss of generality, $\mathfrak{Y} = S$ and $\mathfrak{X} = BG$, for a flat group scheme G over S. Then $\mathfrak{Aut}_\mathfrak{Y} \mathfrak{X} \cong G$, at least locally, and it suffices to prove that there exists a non-empty open subscheme S' of S over which G° is representable. Without loss of generality, S has characteristic $p > 0$. Consider the Frobenius $F : G \to G^{(p)}$. Let G' be the image of F. Clearly, G° is representable if $(G')^\circ$ is. Repeating this process of passing to the image of Frobenius will eventually lead to a generically smooth group scheme, as in the proof of the second devissage lemma, below. □

LEMMA 4.15. *Let $f : \mathfrak{X} \to \mathfrak{Y}$ be a representable étale morphism of algebraic S-stacks of finite type. Then there exists a non-empty open substack $\mathfrak{Y}'$ of $\mathfrak{Y}$ such that the pullback $f' : \mathfrak{X}' \to \mathfrak{Y}'$ of f to $\mathfrak{Y}'$ is finite étale.*

PROOF. Without loss of generality $\mathfrak{X}$ and $\mathfrak{Y}$ are schemes. Then use the fact that f is unramified to reduce to the case that f is separated. Then use Zariski's Main Theorem to conclude. □

For the following two Devissage lemmas let all algebraic stacks considered be of finite type over S.

PROPOSITION 4.16 (First Devissage Lemma). *Let P be a property of morphisms of algebraic S-stacks of finite type. Assume that*

(1) *Given a morphism $f : \mathfrak{X} \to \mathfrak{Y}$ of algebraic S-stacks and an open substack $\mathfrak{U} \subset \mathfrak{X}$ with closed complement $\mathfrak{Z}$, then P holds for f if it holds for $f|\mathfrak{U} : \mathfrak{U} \to \mathfrak{Y}$ and $f|\mathfrak{Z} : \mathfrak{Z} \to \mathfrak{Y}$.*
(2) *If two composable morphisms of algebraic S-stacks satisfy P, then so does their composition.*
(3) *The property P holds for every representable morphism.*
(4) *The property P holds for every gerbe-like morphism whose diagonal has components.*

Then P holds for every morphism of finite type algebraic S-stacks.

PROOF. Let $f : \mathfrak{X} \to \mathfrak{Y}$ be a morphism of algebraic S-stacks. To prove that P holds for f, by (1) and noetherian induction we may replace $\mathfrak{X}$ by some open substack, so we may assume that $\mathfrak{Aut}_\mathfrak{Y} \mathfrak{X} \to \mathfrak{X}$ is flat and has components, by Proposition 4.14. Hence we have a factorization $\mathfrak{X} \to \widetilde{\mathfrak{X}}_\mathfrak{Y} \to \mathfrak{Y}$ of f as in Remark 4.13. Thus by (2) we are done, since $\mathfrak{X} \to \widetilde{\mathfrak{X}}_\mathfrak{Y}$ is gerbe-like with a diagonal having components and $\widetilde{\mathfrak{X}}_\mathfrak{Y} \to \mathfrak{Y}$ is representable. □

PROPOSITION 4.17 (Second Devissage Lemma). *Let P be a property of morphisms of algebraic S-stacks of finite type. Assume that*

(1) *Given a 2-cartesian diagram*

$$\begin{array}{ccc} \mathfrak{X}' & \xrightarrow{f'} & \mathfrak{Y}' \\ v \downarrow & \boxed{2} & \downarrow u \\ \mathfrak{X} & \xrightarrow{f} & \mathfrak{Y} \end{array}$$

of algebraic S-stacks, where u is gerbe-like with Δ_u having components, then P holds for f if it holds for f'.
(2) *Given a morphism $f : \mathfrak{X} \to \mathfrak{Y}$ of algebraic S-stacks and an open substack $\mathfrak{U} \subset \mathfrak{X}$ with closed complement $\mathfrak{Z}$, then P holds for f if it holds for $f|\mathfrak{U} : \mathfrak{U} \to \mathfrak{Y}$ and $f|\mathfrak{Z} : \mathfrak{Z} \to \mathfrak{Y}$.*
(3) *If two composable morphisms of algebraic S-stacks satisfy P, then so does their composition.*
(4) *The property P holds for every representable morphism.*

(5) *The property P holds for the structure morphism $B(G/\mathfrak{X}) \to \mathfrak{X}$, for any smooth group $\mathfrak{X}$-space G with connected fibers over a connected algebraic S-stack $\mathfrak{X}$.*

Then P holds for every morphism of finite type algebraic S-stacks.

PROOF. By the first Devissage we may assume that $f : \mathfrak{X} \to \mathfrak{Y}$ is gerbe-like and Δ_f has components. Factoring f as $\mathfrak{X} \to \overline{\mathfrak{X}}_{\mathfrak{Y}} \to \mathfrak{Y}$, where $\overline{\mathfrak{X}}_{\mathfrak{Y}}$ is as in Proposition 4.11, we may consider two cases, firstly that $\mathfrak{Aut}_{\mathfrak{Y}} \mathfrak{X}$ has connected fibers and secondly that $\mathfrak{Aut}_{\mathfrak{Y}} \mathfrak{X}$ is étale. Making the base change to $\mathfrak{X}$, which is allowed by (1), we may assume that $\mathfrak{X} = B(G/\mathfrak{Y})$, for a flat group space G over $\mathfrak{Y}$, such that G° is representable.

Let us consider the first case, where $\mathfrak{Aut}_{\mathfrak{Y}} \mathfrak{X}$ has connected fibers. In characteristic zero a flat group space with connected fibers is necessarily smooth (see for example Corollaire 3.3.1 in [**10**, Exp. VIIB]). Since smoothness of $G \to \mathfrak{Y}$ is an open property in $\mathfrak{Y}$ (see Proposition 2.5(i) in [**10**, Exp. VIB]), we may thus assume that $\mathfrak{Y}$ is an algebraic stack of characteristic $p > 0$. Let $F : G \to G^{(p)}$ be the Frobenius of G over $\mathfrak{Y}$, $H = \ker F$ and $G' = \operatorname{im} F$. Then $BG \to \mathfrak{Y}$ factors through $BG \to BG'$. This morphism is gerbe-like, linked to H, a group scheme of height ≤ 1. (See Section 7 in [**10**, Exp. VIIA] for the definition of group schemes of height ≤ 1.) So using (1) and (3) we reduce to proving that P holds for $BH \to \mathfrak{Y}$ and $BG' \to \mathfrak{Y}$. Let us assume for the moment that $BH \to \mathfrak{Y}$ has been dealt with, so that we may replace G by G'. Repeating this process of passing to the image of Frobenius, will lead, after a finite number of steps, to a group space $G \to \mathfrak{Y}$ that is generically smooth (see Proposition 8.3 in [**10**, Exp. VIIA]). Again using the fact that smoothness of G is an open property in $\mathfrak{Y}$, we are done, by (2) and (3). Let us now deal with the case where G is a group scheme of height ≤ 1. Since G is finite, we may embed G locally into an abelian scheme (see Theorem A.6 in Chapter III of [**17**]) and thus reduce to (5). This finishes the proof in the first case.

Let us now do the second case, where $\mathfrak{Aut}_{\mathfrak{Y}} \mathfrak{X}$ is étale. By (2) and Lemma 4.15 we may assume that G is finite étale. Letting $\pi : G \to \mathfrak{Y}$ be the structure morphism, we have that $\pi_* \mathcal{O}_G$ is a vector bundle on $\mathfrak{Y}$ and there is a natural monomorphism $G \to GL(\pi_* \mathcal{O}_G)$. Thus $BG \to \mathfrak{Y}$ factors as $BG \to BGL(\pi_* \mathcal{O}_G) \to \mathfrak{Y}$. Now $GL(\pi_* \mathcal{O}_G)$ has connected fibers, so P holds for $BGL(\pi_* \mathcal{O}_G) \to \mathfrak{Y}$ by the first case. On the other hand, $BG \to BGL(\pi_* \mathcal{O}_G)$ is representable. We are done by (4). □

1.3. Universal Homeomorphisms.

DEFINITION 4.18. Let $f : \mathfrak{X} \to \mathfrak{Y}$ be a morphism of algebraic S-stacks which is locally of finite type. First assume that f is representable. We call f a *universal homeomorphism*, if for every scheme $Y \to \mathfrak{Y}$ the pullback $f_Y : \mathfrak{X}_Y \to Y$ is a universal homeomorphism of schemes. Now drop the assumption that f be representable. Then we say that f is a universal homeomorphism if there exists a commutative diagram

$$\begin{array}{ccc} \mathfrak{X}' & & \\ \downarrow & \searrow & \\ \mathfrak{X} & \xrightarrow{f} & \mathfrak{Y}, \end{array}$$

where $\mathfrak{X}' \to \mathfrak{X}$ and $\mathfrak{X}' \to \mathfrak{Y}$ are representable universal homeomorphisms.

NOTE 4.19. A representable universal homeomorphism if finite, radicial and surjective.

EXAMPLE 4.20. Let $\mathfrak{X}$ be an algebraic S-stack of finite type and $G/\mathfrak{X}$ a group space of height ≤ 1. Then $G \to \mathfrak{X}$ is a representable universal homeomorphism and thus $B(G/\mathfrak{X}) \to \mathfrak{X}$ is a universal homeomorphism.

2. The Étale Topos of an Algebraic Stack

2.1. The Flat Topos. Let S be a noetherian scheme. Let $\mathfrak{X}$ be an algebraic S-stack. Consider the site $\mathfrak{S}(\mathfrak{X}_\mathrm{fl})$, defined as follows. The objects of $\mathfrak{S}(\mathfrak{X}_\mathrm{fl})$ are pairs (U, f), where U is an affine S-scheme and $f : U \to \mathfrak{X}$ is a morphism of algebraic S-stacks. For two objects (U, f) and (V, g) a morphism from (U, f) to (V, g) is a pair (ϕ, θ), where $\phi : U \to V$ is a morphism of S-schemes and $\theta : f \to g \circ \phi$ is a 2-isomorphism in the 2-category of algebraic S-stacks. The topology on $\mathfrak{S}(\mathfrak{X}_\mathrm{fl})$ is defined by calling a sieve for (U, f) covering if it contains a finite family $U_i \to U$ of flat morphisms of finite presentation such that $\coprod U_i \to U$ is surjective. We call the topos of sheaves on $\mathfrak{S}(\mathfrak{X}_\mathrm{fl})$ the *fppf-topos* associated to $\mathfrak{X}$ and denote it by $\mathfrak{X}_\mathrm{fl}$.

Consider the following category, which we shall call the *category of $\mathfrak{X}$-spaces*. The objects of ($\mathfrak{X}$-spaces) are pairs $(\mathfrak{Y}, f)$, where $\mathfrak{Y}$ is a (not necessarily algebraic) S-stack and $f : \mathfrak{Y} \to \mathfrak{X}$ is a faithful morphism of S-stacks. Recall that a morphism $f : \mathfrak{Y} \to \mathfrak{X}$ is called *faithful* if for every S-scheme U the morphism of groupoids $f(U) : \mathfrak{Y}(U) \to \mathfrak{X}(U)$ is a faithful functor. For two objects $(\mathfrak{Y}, f)$ and $(\mathfrak{Z}, g)$ of ($\mathfrak{X}$-spaces) a morphism from $(\mathfrak{Y}, f)$ to $(\mathfrak{Z}, g)$ is an equivalence class of pairs (ϕ, θ), where $\phi : \mathfrak{Y} \to \mathfrak{Z}$ is a morphism of algebraic S-stacks and $\theta : f \to g \circ \phi$ is a 2-isomorphism of morphisms of algebraic S-stacks. Here we call two such pairs (ϕ, θ) and (ψ, η) equivalent, if there exists an isomorphism $\xi : \phi \to \psi$ such that $\eta = g(\xi) \circ \theta$. The category of $\mathfrak{X}$-spaces should be thought of as the category of S-stacks that are *relative sheaves* over $\mathfrak{X}$.

PROPOSITION 4.21. *The category of $\mathfrak{X}$-spaces is canonically isomorphic to $\mathfrak{X}_\mathrm{fl}$.*

PROOF. The site $\mathfrak{S}(\mathfrak{X}_\mathrm{fl})$ is naturally a full subcategory of ($\mathfrak{X}$-spaces). The category of $\mathfrak{X}$-spaces is a topos. Then conclude using the comparison lemma Théorème 4.1 from [**2**, Exp. III]. □

DEFINITION 4.22. Let $\mathfrak{X}_\mathrm{alg}$ be the full subcategory of ($\mathfrak{X}$-spaces) consisting of objects $(\mathfrak{Y}, f)$ such that $\mathfrak{Y}$ is an *algebraic* S-stack. We call $\mathfrak{X}_\mathrm{alg}$ the category of *algebraic $\mathfrak{X}$-spaces*.

2.2. The Étale Topos. Fix an algebraic S-stack $\mathfrak{X}$. Let $\mathfrak{X}_{\text{ét}}$ be the full subcategory of $\mathfrak{X}_{\text{alg}}$, whose objects are algebraic S-stacks that are étale over $\mathfrak{X}$.

LEMMA 4.23. *The category $\mathfrak{X}_{\text{ét}}$ is a topos. The open subtopoi of $\mathfrak{X}_{\text{ét}}$ correspond bijectively to the open substacks of $\mathfrak{X}$. The points of $\mathfrak{X}$ give rise to a conservative set of points of $\mathfrak{X}_{\text{ét}}$. If $\mathfrak{X}$ is of finite type, then $\mathfrak{X}_{\text{ét}}$ is noetherian. Every morphism $f : \mathfrak{X} \to \mathfrak{Y}$ of algebraic S-stacks induces a morphism $f : \mathfrak{X}_{\text{ét}} \to \mathfrak{Y}_{\text{ét}}$ of the associated étale topoi.*

PROOF. The fact that $\mathfrak{X}_{\text{ét}}$ is a topos is standard, using our requirement that all algebraic stacks be locally noetherian. The fact that étale morphisms are open implies the claim concerning open subtopoi of $\mathfrak{X}_{\text{ét}}$. The claim about points is implied by the fact that a radicial étale epimorphism is an isomorphism. The finite type objects of $\mathfrak{X}_{\text{ét}}$ form a generating family consisting of noetherian objects. Thus $\mathfrak{X}_{\text{ét}}$ is noetherian, if $\mathfrak{X}$ is of finite type. The claim about morphisms follows from the fact that pullback commutes with fibered products and takes epimorphisms to epimorphisms. □

LEMMA 4.24. *Let $i : \mathfrak{Z} \to \mathfrak{X}$ be a closed substack of $\mathfrak{X}$ and $j : \mathfrak{U} \to \mathfrak{X}$ the open complement of $\mathfrak{Z}$. Consider $\mathfrak{U}$ as an object of $\mathfrak{X}_{\text{ét}}$. Then $j_* : \mathfrak{U}_{\text{ét}} \to \mathfrak{X}_{\text{ét}}$ identifies $\mathfrak{U}_{\text{ét}}$ with the open subtopos of $\mathfrak{X}_{\text{ét}}$ defined by $\mathfrak{U}$. The functor $i_* : \mathfrak{Z}_{\text{ét}} \to \mathfrak{X}_{\text{ét}}$ identifies $\mathfrak{Z}_{\text{ét}}$ with the closed complement of this open subtopos of $\mathfrak{X}_{\text{ét}}$.*

PROOF. The claim about j_* is trivial. To prove the claim about i_* we need to prove two facts. Firstly, if F is an étale sheaf on $\mathfrak{Z}$ then $i^* i_* F \to F$ is an isomorphism. Secondly, if G is an étale sheaf on $\mathfrak{X}$ such that $G \times \mathfrak{U} = \mathfrak{U}$, then $G \to i_* i^* G$ is an isomorphism. Both of these statements are easily seen to be local in $\mathfrak{X}$, so we may assume that $\mathfrak{X}$ is a scheme. The first is a formal consequence of the fact that $\mathfrak{Z} \times_{\mathfrak{X}} \mathfrak{Z} = \mathfrak{Z}$. The second is a formal consequence of the first and the following fact: Let $\phi : F \to G$ be a morphism of étale $\mathfrak{X}$-sheaves. If $\phi|\mathfrak{U}$ and $\phi|\mathfrak{Z}$ are isomorphisms, then so is ϕ. This fact follows from the fact that the points of $\mathfrak{X}$ form a conservative family of points for $\mathfrak{X}_{\text{ét}}$. □

Note that, in particular, a nilpotent closed immersion induces an isomorphism on the étale topoi.

COROLLARY 4.25. *Let $k : \mathfrak{V} \to \mathfrak{X}$ be a locally closed immersion of algebraic S-stacks. Then the induced morphism $k : \mathfrak{V}_{\text{ét}} \to \mathfrak{X}_{\text{ét}}$ is a locally closed immersion of topoi. If $k' : \mathfrak{V}' \to \mathfrak{X}'$ is a base change of k given by $u : \mathfrak{X}' \to \mathfrak{X}$, then $k'_*(\mathfrak{V}'_{\text{ét}})$ is the preimage of $k_*(\mathfrak{V}_{\text{ét}})$ under the morphism of topoi $u : \mathfrak{X}'_{\text{ét}} \to \mathfrak{X}_{\text{ét}}$.*

REMARK 4.26. If $f : \mathfrak{X} \to \mathfrak{Y}$ is a universal homeomorphism, then $f : \mathfrak{X}_{\text{ét}} \to \mathfrak{Y}_{\text{ét}}$ is an isomorphism of topoi. In other words, $f_* : \mathfrak{X}_{\text{ét}} \to \mathfrak{Y}_{\text{ét}}$ is an equivalence of categories with quasi-inverse f^*.

PROPOSITION 4.27. *Let $f : \mathfrak{Y} \to \mathfrak{X}$ be a representable étale morphism of algebraic stacks. Then f represents an lcc-sheaf on $\mathfrak{X}_{ét}$ if and only if f is finite.*

PROOF. If $\mathfrak{Y}$ is an lcc sheaf then f is finite, since finiteness is local with respect to the étale topology. Conversely, assume that f is finite. Let n be the degree of f. Making the base change from $\mathfrak{X}$ to $\mathfrak{Y}$ we may assume that f has a section. Then $\mathfrak{Y} = \mathfrak{X} \coprod \mathfrak{Y}'$, where $\mathfrak{Y}'$ is finite étale of degree $n - 1$ over $\mathfrak{X}$. By induction $\mathfrak{Y}'$ is lcc, thus $\mathfrak{Y}$ is lcc, too. □

PROPOSITION 4.28. *Let $\mathfrak{X}$ be an algebraic S-stack of finite type. If F is a sheaf on $\mathfrak{X}_{ét}$ then F is constructible if and only if F is represented by an étale morphism of finite type $f : \mathfrak{Y} \to \mathfrak{X}$ of algebraic S-stacks.*

PROOF. If F is represented by a finite type morphism $f : \mathfrak{Y} \to \mathfrak{X}$, then F is constructible by Proposition 4.27 and Lemma 4.15. For the converse, let $f : \mathfrak{Y} \to \mathfrak{X}$ be representable, étale such that the induced sheaf on $\mathfrak{X}_{ét}$ is constructible. To prove that f is of finite type it suffices to check that $\mathfrak{Y}$ is quasi compact. But this is easy. □

PROPOSITION 4.29. *Let $f : \mathfrak{X} \to \mathfrak{Y}$ be a diagonally connected gerbe-like morphism of algebraic S-stacks of finite type. Then f induces an isomorphism of étale topoi $\mathfrak{X}_{ét} \xrightarrow{\sim} \mathfrak{Y}_{ét}$.*

PROOF. By faithfully flat descent, we may assume without loss of generality that f admits a section and $\mathfrak{Y} = S$ is a scheme. Then $\mathfrak{X}$ is a neutral gerbe over S. So there is an S-group G with connected fibers such that $\mathfrak{X} = BG$. The category of representable étale BG-stacks is equivalent to the category of étale S-schemes with G-action. But the only way a connected group can act on an étale scheme is trivially. □

COROLLARY 4.30. *Let $\mathfrak{X}$ be an algebraic S-stack such that $\mathfrak{Aut}\,\mathfrak{X} \to \mathfrak{X}$ is flat and has components. Then the morphism of algebraic S-stacks $\mathfrak{X} \to \overline{\mathfrak{X}}$ induces an isomorphism of associated étale topoi $\mathfrak{X}_{ét} \to \overline{\mathfrak{X}}_{ét}$. Here $\overline{\mathfrak{X}}$ is the associated Deligne-Mumford stack (see Remark 4.12).*

PROOF. By construction, $\mathfrak{X} \to \overline{\mathfrak{X}}$ satisfies the requirements of the proposition. □

3. The Smooth Topos

Let $\mathfrak{X}$ continue to denote an algebraic S-stack. Let $\mathfrak{Y}$ be another algebraic S-stack and $\pi : \mathfrak{X} \to \mathfrak{Y}$ a fixed morphism, which we assume to be smooth and representable. We define the *smooth site of $\mathfrak{X}$ relative to $\mathfrak{Y}$*, denoted $\mathfrak{S}(\mathfrak{X}_{\mathfrak{Y}\text{-sm}})$, as follows.

DEFINITION 4.31. The underlying category of $\mathfrak{S}(\mathfrak{X}_{\mathfrak{Y}\text{-sm}})$ is the full subcategory of $\mathfrak{X}_{alg}$ consisting of algebraic $\mathfrak{X}$-spaces that are smooth over $\mathfrak{Y}$. A sieve is called covering if it contains a finite number of smooth morphisms whose images cover. We denote by $\mathfrak{X}_{\mathfrak{Y}\text{-sm}}$ the corresponding topos of sheaves and call it the *smooth topos of* $\mathfrak{X}$ *relative to* $\mathfrak{Y}$. If $\mathfrak{X} = \mathfrak{Y}$, then we set $\mathfrak{X}_{sm} = \mathfrak{X}_{x\text{-}sm}$ and call it the *smooth topos of* $\mathfrak{X}$.

NOTE 4.32. The topos $\mathfrak{X}_{\mathfrak{Y}\text{-}sm}$ may be considered as the induced topos $(\mathfrak{Y}_{sm})_{/\mathfrak{X}}$, where $\mathfrak{X}$ is considered as a (representable) sheaf on $\mathfrak{S}(\mathfrak{Y}_{sm})$.

REMARK 4.33. The inclusion functor $\mathfrak{S}(\mathfrak{X}_{\mathfrak{Y}\text{-}sm}) \to \mathfrak{X}_{fl}$ is clearly continuous. Thus it induces a pseudo-morphism of topoi $v : \mathfrak{X}_{fl} \to \mathfrak{X}_{\mathfrak{Y}\text{-}sm}$, such that v^* extends the embedding and v_* associates to an $\mathfrak{X}$-space the induced sheaf on $\mathfrak{S}(\mathfrak{X}_{\mathfrak{Y}\text{-}sm})$. Note that v_* commutes with fibered products but is not fully faithful. Since $\mathfrak{S}(\mathfrak{X}_{\mathfrak{Y}\text{-}sm})$ is not closed under fibered products, v is not a morphism of topoi.

DEFINITION 4.34. A morphism $\phi : F \to G$ in $\mathfrak{X}_{\mathfrak{Y}\text{-}sm}$ is called *étale*, if for every $U \in \text{ob}\,\mathfrak{S}(\mathfrak{X}_{\mathfrak{Y}\text{-}sm})$ and every $s \in G(U)$ we have that $F \times_{G,s} U$ is representable (by an object of $\mathfrak{S}(\mathfrak{X}_{\mathfrak{Y}\text{-}sm})$) and $F \times_{G,s} U \to U$ is an étale morphism of algebraic S-stacks.

NOTE 4.35. For $\phi : F \to G$ to be étale it suffices that there exists and epimorphism $U \to G$, such that $F \times_{G,s} U \to U$ is étale.

If $\phi : F \to G$ is an étale epimorphism, then G is representable if and only if F is representable.

PROPOSITION 4.36. *The set of étale morphisms is an e-structure (see Definition 3.60) on* $\mathfrak{X}_{\mathfrak{Y}\text{-}sm}$.

Thus for every object U of $\mathfrak{X}_{\mathfrak{Y}\text{-}sm}$, letting $\overline{U}$ be the category of sheaves that are étale over U, we have a c-topos $(U, \overline{U})$. If U is in $\mathfrak{S}(\mathfrak{X}_{\mathfrak{Y}\text{-}sm})$, represented by the algebraic stack $\mathfrak{U}$, then we have $(U, \overline{U}) = (\mathfrak{U}_{\mathfrak{Y}\text{-}sm}, \mathfrak{U}_{ét})$.

PROPOSITION 4.37. *If* $\mathfrak{X}$ *is a Deligne-Mumford stack, then* $(\mathfrak{X}_{\mathfrak{Y}\text{-}sm}, \mathfrak{X}_{ét})$ *is a trivial c-topos.*

PROOF. We need to show that $\pi^* : \mathfrak{X}_{ét} \to \mathfrak{X}_{\mathfrak{Y}\text{-}sm}$ is cocontinuous. This reduces to proving that for every smooth representable epimorphism $\mathfrak{X}' \to \mathfrak{X}$ there exists an étale surjection $\widetilde{\mathfrak{X}} \to \mathfrak{X}$ over which $\mathfrak{X}' \to \mathfrak{X}$ has a section. This follows from the corresponding result for schemes using the fact that Deligne-Mumford stacks have étale presentations. □

REMARK 4.38. Let $\mathfrak{X}$ be an algebraic S-stack of finite type and A a discrete valuation ring. Then the c-topos $(\mathfrak{X}_{sm}, \mathfrak{X}_{ét})$ is noetherian by Lemma 4.23. So as in Definition 3.46 we get the category $\mathbb{D}_c(\mathfrak{X}_{sm}, A)$ of constructible A-complexes on $\mathfrak{X}_{sm}$. It is an A-t-category with heart $\text{Mod}_c(\mathfrak{X}_{ét}, A)$, the category of constructible A-sheaves on $\mathfrak{X}_{ét}$.

REMARK 4.39. Let us see what happens if we change the topology on $\mathfrak{X}$. Consider a commutative diagram

$$\begin{array}{ccc} \mathfrak{X} & \xrightarrow{\pi} & \mathfrak{Y} \\ {\scriptstyle \pi'} \searrow & & \downarrow g \\ & & \mathfrak{Y}', \end{array}$$

where π and π' are smooth and representable (so that g is smooth, too). We have a natural embedding $\mathfrak{S}(\mathfrak{X}_{\mathfrak{Y}\text{-sm}}) \to \mathfrak{S}(\mathfrak{X}_{\mathfrak{Y}'\text{-sm}})$, which is trivially at the same time time continuous and cocontinuous. Thus we get an induced morphism of topoi

$$u : \mathfrak{X}_{\mathfrak{Y}\text{-sm}} \to \mathfrak{X}_{\mathfrak{Y}'\text{-sm}},$$

such that u^* has a left adjoint $u_! : \mathfrak{X}_{\mathfrak{Y}\text{-sm}} \to \mathfrak{X}_{\mathfrak{Y}'\text{-sm}}$, which extends the embedding $\mathfrak{S}(\mathfrak{X}_{\mathfrak{Y}\text{-sm}}) \to \mathfrak{S}(\mathfrak{X}_{\mathfrak{Y}'\text{-sm}})$. The functor u^* restricts a sheaf from $\mathfrak{S}(\mathfrak{X}_{\mathfrak{Y}'\text{-sm}})$ to $\mathfrak{S}(\mathfrak{X}_{\mathfrak{Y}\text{-sm}})$.

Now consider 2-commutative diagrams of algebraic S-stacks

$$\begin{array}{ccc} \mathfrak{X} & \xrightarrow{f} & \mathfrak{X}' \\ \pi \downarrow & & \downarrow \pi' \\ \mathfrak{Y} & \xrightarrow{g} & \mathfrak{Y}' \end{array} \qquad (5)$$

where π and π' are smooth and representable. We will study to what extent f induces a morphism of topoi from $\mathfrak{X}_{\mathfrak{Y}\text{-sm}}$ to $\mathfrak{X}'_{\mathfrak{Y}'\text{-sm}}$.

First, let us consider the case where (5) is 2-cartesian. Pullback via f defines a functor

$$f^* : \mathfrak{S}(\mathfrak{X}'_{\mathfrak{Y}'\text{-sm}}) \longrightarrow \mathfrak{S}(\mathfrak{X}_{\mathfrak{Y}\text{-sm}}).$$

This functor is continuous, in other words induces a functor

$$f_* : \mathfrak{X}_{\mathfrak{Y}\text{-sm}} \longrightarrow \mathfrak{X}'_{\mathfrak{Y}'\text{-sm}},$$

given by $f_*F(U) = F(f^*U)$, for a sheaf $F \in \text{ob}\,\mathfrak{X}_{\mathfrak{Y}\text{-sm}}$ and an object $U \in \text{ob}\,\mathfrak{S}(\mathfrak{X}'_{\mathfrak{Y}'\text{-sm}})$. By Proposition 1.2 in [**2**, Exp. III] f^* extends thus to a functor

$$f^* : \mathfrak{X}'_{\mathfrak{Y}'\text{-sm}} \longrightarrow \mathfrak{X}_{\mathfrak{Y}\text{-sm}},$$

which is a left adjoint of f_*.

Let us now consider the case where in (5) we have $\mathfrak{Y} = \mathfrak{Y}'$. Then f may be considered as a morphism in $\mathfrak{S}(\mathfrak{Y}_{\text{sm}})$ or $\mathfrak{S}(\mathfrak{X}'_{\mathfrak{Y}'\text{-sm}})$. (Note that f is representable.) So f induces a morphism of topoi $f : \mathfrak{X}_{\mathfrak{Y}\text{-sm}} \to \mathfrak{X}'_{\mathfrak{Y}\text{-sm}}$.

Finally, for treating the general case of (5), we may write f as a composition of the second case followed by the first. We get a pair of adjoint functors (f^*, f_*) between $\mathfrak{X}_{\mathfrak{Y}\text{-sm}}$ and $\mathfrak{X}'_{\mathfrak{Y}'\text{-sm}}$. In other words, we get a pseudo-morphism of topoi $f_{\text{sm}} : \mathfrak{X}_{\mathfrak{Y}\text{-sm}} \to \mathfrak{X}'_{\mathfrak{Y}'\text{-sm}}$ (see Remark 3.32).

NOTE 4.40. If in Diagram (5) $\mathfrak{X} \to \mathfrak{Y}'$ is smooth and representable, then the pseudo-morphism $f : \mathfrak{X}_{\mathfrak{Y}\text{-sm}} \to \mathfrak{X}'_{\mathfrak{Y}'\text{-sm}}$ may be factored as $\mathfrak{X}_{\mathfrak{Y}\text{-sm}} \to \mathfrak{X}_{\mathfrak{Y}'\text{-sm}} \to \mathfrak{X}'_{\mathfrak{Y}'\text{-sm}}$, where $\mathfrak{X}_{\mathfrak{Y}\text{-sm}} \to \mathfrak{X}_{\mathfrak{Y}'\text{-sm}}$ is the morphism defined in Remark 4.39. In particular, $f : \mathfrak{X}_{\mathfrak{Y}\text{-sm}} \to \mathfrak{X}'_{\mathfrak{Y}'\text{-sm}}$ is a morphism of topoi.

PROPOSITION 4.41. *If $f : \mathfrak{X} \to \mathfrak{X}'$ is smooth, then f^* is exact.*

PROOF. If f is representable, this follows from Note 4.40. For the general case, let $X \to \mathfrak{X}$ be a presentation of $\mathfrak{X}$. Then the morphism $u : X_{\mathfrak{Y}\text{-sm}} \to \mathfrak{X}_{\mathfrak{Y}\text{-sm}}$ has the property that if u^* of a diagram is cartesian, the original diagram was cartesian. This allows us to replace $\mathfrak{X}_{\mathfrak{Y}\text{-sm}} \to \mathfrak{X}'_{\mathfrak{Y}'\text{-sm}}$ by $X_{\mathfrak{Y}\text{-sm}} \to \mathfrak{X}'_{\mathfrak{Y}'\text{-sm}}$. Then we are in the representable case. □

WARNING 4.42. In general, f^* is not exact. Consider, for example, the case where $\mathfrak{X} = \mathfrak{Y}$, $\mathfrak{X}' = \mathfrak{Y}'$ and $f : \mathfrak{X} \to \mathfrak{X}'$ is a closed immersion of smooth varieties over $S = \operatorname{Spec} k$, k an algebraically closed field. Let $s \neq 0$ be a regular function on $\mathfrak{X}'$, vanishing along $\mathfrak{X}$. Consider s as a homomorphism of group schemes $s : \mathbb{A}^1_{\mathfrak{X}'} \to \mathbb{A}^1_{\mathfrak{X}'}$. Then we may, in fact, consider s as a homomorphism of representable abelian sheaves on $\mathfrak{S}(\mathfrak{X}'_{\mathrm{sm}})$. We have $\ker(s) = 0$. Now pulling back to $\mathfrak{X}$, the morphism $f^*(s) : \mathbb{A}^1_{\mathfrak{X}} \to \mathbb{A}^1_{\mathfrak{X}}$ is the zero homomorphism, and so $\ker(f^*(s)) = \mathbb{A}^1_{\mathfrak{X}}$.

4. The Simplicial Approach

Let $\mathfrak{X}$ be an algebraic S-stack and $X \to \mathfrak{X}$ a presentation of $\mathfrak{X}$. Then we construct a simplicial algebraic space $X_\bullet$ by setting $X_{\Delta_n} = X_n = \underbrace{X \times_\mathfrak{X} X \ldots \times_\mathfrak{X} X}_{n+1}$ and by assigning projections to the face maps and diagonals to the degeneracy maps, analogously to the construction of the Čech nerve of a one-element covering. The simplicial algebraic space $X_\bullet$ gives rise to a simplicial topos $X_{\bullet\mathrm{\acute{e}t}}$, whose fiber over Δ_n is the étale topos of X_n. The associated total topos (see Section 1) is called the *étale topos of $X_\bullet$*, and is denoted $\operatorname{top}(X_{\bullet\mathrm{\acute{e}t}})$ or by abuse of notation $X_{\bullet\mathrm{\acute{e}t}}$, if no confusion is likely to arise.

By descent theory we have that $X_{\bullet\mathrm{\acute{e}t},\mathrm{cart}} = \mathfrak{X}_{\mathrm{\acute{e}t}}$ and so $\mathfrak{X}_{\mathrm{\acute{e}t}}$ defines a c-structure on $\operatorname{top}(X_{\bullet\mathrm{\acute{e}t}})$, which is noetherian if $\mathfrak{X}$ is of finite type (see Lemma 4.23).

As in Example 3.3 we also get an associated simplicial topos $X^{\mathbb{N}}_{\bullet\,\mathrm{\acute{e}t}}$, whose fiber over Δ_n is $X^{\mathbb{N}}_{n\,\mathrm{\acute{e}t}}$, the category of projective systems of sheaves on $X_{n\,\mathrm{\acute{e}t}}$. The associated total topos $\operatorname{top}(X^{\mathbb{N}}_{\bullet\,\mathrm{\acute{e}t}})$ is canonically equivalent to $\operatorname{top}(X_{\bullet\mathrm{\acute{e}t}})^{\mathbb{N}}$. The associated topos of cartesian objects $X^{\mathbb{N}}_{\bullet\,\mathrm{\acute{e}t},\mathrm{cart}}$ is canonically equivalent to $\mathfrak{X}^{\mathbb{N}}_{\mathrm{\acute{e}t}}$. So we get a c-topos $(\operatorname{top}(X^{\mathbb{N}}_{\bullet\,\mathrm{\acute{e}t}}), \mathfrak{X}^{\mathbb{N}}_{\mathrm{\acute{e}t}})$ as in Example 3.36.

Now let $f : \mathfrak{X} \to \mathfrak{Y}$ be a morphism of algebraic S-stacks and $f_0 : X \to Y$ a morphism of presentations $X \to \mathfrak{X}$ and $Y \to \mathfrak{Y}$ such that the diagram

$$\begin{array}{ccc} X & \xrightarrow{f_0} & Y \\ \downarrow & & \downarrow \\ \mathfrak{X} & \xrightarrow{f} & \mathfrak{Y} \end{array} \qquad (6)$$

commutes. We get an induced morphism $f_\bullet : X_\bullet \to Y_\bullet$ of simplicial algebraic spaces, an induced morphism of simplicial topoi $f_\bullet : X_{\bullet\mathrm{\acute{e}t}} \to Y_{\bullet\mathrm{\acute{e}t}}$ and an induced morphism of the associated total topoi $\operatorname{top}(f_\bullet) : \operatorname{top}(X_{\bullet\mathrm{\acute{e}t}}) \to \operatorname{top}(Y_{\bullet\mathrm{\acute{e}t}})$. Clearly, $f_\bullet$ induces a morphism of the first kind of c-topoi

$$f_\bullet : (\operatorname{top}(X_{\bullet\mathrm{\acute{e}t}}), \mathfrak{X}_{\mathrm{\acute{e}t}}) \longrightarrow (\operatorname{top}(Y_{\bullet\mathrm{\acute{e}t}}), \mathfrak{Y}_{\mathrm{\acute{e}t}}).$$

We also get an induced morphism of simplicial topoi $f^{\mathbb{N}}_\bullet : X^{\mathbb{N}}_{\bullet\,\mathrm{\acute{e}t}} \to Y^{\mathbb{N}}_{\bullet\,\mathrm{\acute{e}t}}$ inducing a morphism of total topoi $\operatorname{top}(f^{\mathbb{N}}_\bullet) : \operatorname{top}(X^{\mathbb{N}}_{\bullet\,\mathrm{\acute{e}t}}) \to \operatorname{top}(Y^{\mathbb{N}}_{\bullet\,\mathrm{\acute{e}t}})$. Clearly, we have $\operatorname{top}(f^{\mathbb{N}}_\bullet) = \operatorname{top}(f_\bullet)^{\mathbb{N}}$, so we may just write $f^{\mathbb{N}}_\bullet$ instead.

This morphism induces a morphism of the first kind of c-topoi

$$f_\bullet^\mathbb{N} : (\mathrm{top}(X_{\bullet\,\mathrm{\acute{e}t}}^\mathbb{N}), \mathfrak{X}_{\mathrm{\acute{e}t}}^\mathbb{N}) \longrightarrow (\mathrm{top}(Y_{\bullet\,\mathrm{\acute{e}t}}^\mathbb{N}), \mathfrak{Y}_{\mathrm{\acute{e}t}}^\mathbb{N}).$$

Now let $f : \mathfrak{X} \to \mathfrak{Y}$ be representable and choose $X \to \mathfrak{X}$ such that the diagram (6) is 2-cartesian. Let A' be a ring whose characteristic is invertible on S. Then

$$f_\bullet : (\mathrm{top}(X_{\bullet\mathrm{\acute{e}t}}), \mathfrak{X}_{\mathrm{\acute{e}t}}) \longrightarrow (\mathrm{top}(Y_{\bullet\mathrm{\acute{e}t}}), \mathfrak{Y}_{\mathrm{\acute{e}t}})$$

is of the second kind with respect to A'. This follows immediately from the smooth base change theorem.

Now let A be a discrete valuation ring whose residue characteristic is invertible on S.

REMARK 4.43. Let $\Lambda = (A/\ell^{n+1})_{n\in\mathbb{N}}$, which we may consider as a sheaf of A-algebras on $\mathfrak{X}_{\mathrm{\acute{e}t}}^\mathbb{N}$ and $\mathfrak{Y}_{\mathrm{\acute{e}t}}^\mathbb{N}$. Then

$$f_\bullet^\mathbb{N} : (\mathrm{top}(X_{\bullet\,\mathrm{\acute{e}t}}^\mathbb{N}), \mathfrak{X}_{\mathrm{\acute{e}t}}^\mathbb{N}) \longrightarrow (\mathrm{top}(Y_{\bullet\,\mathrm{\acute{e}t}}^\mathbb{N}), \mathfrak{Y}_{\mathrm{\acute{e}t}}^\mathbb{N})$$

is of the second kind with respect to Λ.

Assume now that S is of finite type over a regular noetherian scheme of dimension zero or one.

PROPOSITION 4.44. *Let $f : \mathfrak{X} \to \mathfrak{Y}$ be a representable morphism of finite type algebraic S-stacks. Then*

$$f_\bullet : (\mathrm{top}(X_{\bullet\,\mathit{\acute{e}t}}), \mathfrak{X}_{\mathit{\acute{e}t}}) \longrightarrow (\mathrm{top}(Y_{\bullet\,\mathit{\acute{e}t}}), \mathfrak{Y}_{\mathit{\acute{e}t}})$$

induces a functor $Rf_{\bullet*} : D_c^+(X_{\bullet\,\mathit{\acute{e}t}}, A) \to D_c^+(Y_{\bullet\,\mathit{\acute{e}t}}, A)$ (still assuming that X is the pullback of Y).

PROOF. Let F be a constructible sheaf of A-modules on $\mathfrak{X}_{\mathrm{\acute{e}t}}$. Then there exists an n such that F is a sheaf of A/ℓ^{n+1}-modules. Hence $R^q f_{\bullet*} F^\bullet$ is a cartesian object of $\mathrm{top}(Y_{\bullet\,\mathrm{\acute{e}t}})$. Now F is represented by a representable étale morphism of finite type $F \to \mathfrak{X}$ of algebraic S-stacks. Fix for the moment a $p \geq 0$ and denote by F^p the pullback of F to X_p. There exists a q_0 such that for all $q \geq q_0$ we have $R^q f_{p*} F^p = 0$. By Deligne's finiteness theorem (Théorème 1.1 of [**8**, Th. finitude.]) $R^q f_{p*} F^p$ is constructible for all p,q, and thus represented by an étale Y_p-scheme of finite type. Fix q. Denote by $R^q f_* F$ the finite type stack over $\mathfrak{Y}$ defined by $R^q f_{\bullet*} F^\bullet$. (N.B. This is *not* the derived functor of $f_{\mathrm{\acute{e}t}*} : \mathrm{Mod}(\mathfrak{X}_{\mathrm{\acute{e}t}}, A) \to \mathrm{Mod}(\mathfrak{Y}_{\mathrm{\acute{e}t}}, A)$.) The $\mathfrak{Y}$-space $R^q f_* F$ is of finite type, hence constructible (see Proposition 4.28). The claim follows. $\square$

4.1. The d-structures defined by the simplicial c-topos. Fix a presentation $X \to \mathfrak{X}$ of the algebraic S-stack $\mathfrak{X}$. For any locally closed substack $\mathfrak{V}$ of $\mathfrak{X}$ denote by $V \to \mathfrak{V}$ the induced presentation. By Corollaries 4.25 and 3.10 we have the following. If $\mathfrak{X} = \mathfrak{U} \cup \mathfrak{Z}$ is the disjoint union of an open substack $\mathfrak{U}$ and a closed substack $\mathfrak{Z}$, then

$$\begin{array}{ccccc} \mathrm{top}(U_{\bullet\,\mathrm{\acute{e}t}}) & \longrightarrow & \mathrm{top}(X_{\bullet\,\mathrm{\acute{e}t}}) & \longleftarrow & \mathrm{top}(Z_{\bullet\,\mathrm{\acute{e}t}}) \\ \downarrow & \boxed{2} & \downarrow & \boxed{2} & \downarrow \\ \mathfrak{U}_{\mathrm{\acute{e}t}} & \longrightarrow & \mathfrak{X}_{\mathrm{\acute{e}t}} & \longleftarrow & \mathfrak{Z}_{\mathrm{\acute{e}t}} \end{array}$$

are 2-cartesian diagrams. For every locally closed immersion $\mathfrak{V} \to \mathfrak{X}$ we have a 2-cartesian diagram

$$\begin{array}{ccc} \mathrm{top}(V_{\bullet\,\mathrm{\acute{e}t}}) & \longrightarrow & \mathrm{top}(X_{\bullet\,\mathrm{\acute{e}t}}) \\ \downarrow & & \downarrow \\ \mathfrak{V}_{\mathrm{\acute{e}t}} & \longrightarrow & \mathfrak{X}_{\mathrm{\acute{e}t}}. \end{array}$$

So $(\mathrm{top}(V_{\bullet\,\mathrm{\acute{e}t}}), \mathfrak{V}_{\mathrm{\acute{e}t}})$ is the induced c-topos (see Definition 3.52) over the locally closed subtopos $\mathfrak{V}_{\mathrm{\acute{e}t}} \subset \mathfrak{X}_{\mathrm{\acute{e}t}}$.

Let A be a discrete valuation ring whose residue characteristic is invertible on S, which we assume to be of finite type over some noetherian regular scheme of dimension zero or one. By Propositions 4.44 and 3.57 (and the note following it) we get a tractable A-cd-structure

$$\mathfrak{V} \longmapsto D^+_{\mathrm{lcc}}(V_{\bullet\,\mathrm{\acute{e}t}}, A) \subset D^+_c(V_{\bullet\,\mathrm{\acute{e}t}}, A)$$

over the topological space $|\mathfrak{X}|$.

COROLLARY 4.45. *In the situation of Proposition 4.44 the morphism* $f_\bullet : (\mathrm{top}(X_{\bullet\,\mathrm{\acute{e}t}}), \mathfrak{X}_{\mathrm{\acute{e}t}}) \to (\mathrm{top}(Y_{\bullet\,\mathrm{\acute{e}t}}), \mathfrak{Y}_{\mathrm{\acute{e}t}})$ *is a morphism of c-topoi of the third kind with respect to* A.

PROOF. Proposition 4.44 says that f induces a tractable morphism of cd-structures. Apply Lemma 2.16. □

By Remark 4.43 and Example 3.56, we get an A-d-structure

$$\mathfrak{V} \longmapsto D^+_{\mathrm{cart}}(V^{\mathbb{N}}_{\bullet\,\mathrm{\acute{e}t}}, \Lambda)$$

over the topological space $|\mathfrak{X}|$, which induces an A-d-structure

$$\mathfrak{V} \longmapsto \mathbb{D}^+_c(V_{\bullet\,\mathrm{\acute{e}t}}, \Lambda)$$

by Proposition 3.59.

5. The Smooth Approach

REMARK 4.46. Let us consider the topos $\mathfrak{X}_{\mathfrak{Y}\text{-sm}}$ endowed with its étale e-structure (see Proposition 4.36). Let $\mathcal{U}$ be the set of presentations of $\mathfrak{X}$, considered as a family of objects of $\mathfrak{S}(\mathfrak{X}_{\mathfrak{Y}\text{-sm}})$. By Proposition 4.37 the family $\mathcal{U}$ satisfies the conditions required of $\mathcal{U}$ in Section 6. So by Proposition 3.63 we get for any presentation $X \to \mathfrak{X}$ of $\mathfrak{X}$ that

$$\sigma_* \pi^* : D^+_{\mathrm{\acute{e}t}}(\mathfrak{X}_{\mathfrak{Y}\text{-sm}}, \Lambda) \longrightarrow D^+_{\mathrm{cart}}(X_{\bullet\,\mathrm{\acute{e}t}}, \Lambda)$$

is an equivalence of categories, where Λ is any sheaf of rings on $\mathfrak{X}_{\mathrm{\acute{e}t}}$. Note that σ_* is exact, so it is not necessary to derive $\sigma_* \pi^*$.

We also get from Corollary 3.64 that

$$\sigma_* \pi^* : \mathbb{D}^+_c(\mathfrak{X}_{\mathfrak{Y}\text{-sm}}, A) \longrightarrow \mathbb{D}^+_c(X_{\bullet\,\mathrm{\acute{e}t}}, A)$$

is an equivalence of categories, for any discrete valuation ring A.

NOTE 4.47. Since we clearly have a morphism $u^* : D_{\text{ét}}^+(\mathfrak{X}_{\mathfrak{Y}\text{-sm}}, \Lambda) \to D_{\text{ét}}^+(\mathfrak{X}_{\text{sm}}, \Lambda)$, we get as a Corollary of Remark 4.46 that it is an equivalence of categories. So for the definition of $D_{\text{ét}}^+(\mathfrak{X}_{\text{sm}}, \Lambda)$ it is immaterial with which smooth topos we work. This is not the case, however, for the definition of direct image functors. As we shall see, to define a direct image functor

$$Rf_* : D_{\text{ét}}^+(\mathfrak{X}_{\text{sm}}, \Lambda) \longrightarrow D_{\text{ét}}^+(\mathfrak{X}'_{\text{sm}}, \Lambda)$$

associated to a morphism of algebraic S-stacks $f : \mathfrak{X} \to \mathfrak{X}'$, it is essential that, at least over $\mathfrak{X}'$, we work with the absolute smooth topos, and not a relative one.

Now fix a ring A', whose characteristic is invertible on S. Thus the smooth base change theorem holds for étale sheaves of A'-modules over Deligne-Mumford stacks over S.

LEMMA 4.48. *Let $f : \mathfrak{X} \to \mathfrak{Y}$ be a representable morphism of algebraic S-stacks. Let $U \to V$ be a morphism of smooth $\mathfrak{Y}$-schemes. Using notations as in the diagram*

$$\begin{array}{ccc} Z & \xrightarrow{h} & U \\ v \downarrow & \boxed{2} & \downarrow u \\ W & \xrightarrow{g} & V \\ t \downarrow & \boxed{2} & \downarrow \\ \mathfrak{X} & \xrightarrow{f} & \mathfrak{Y} \end{array}$$

we have for any étale sheaf of A'-modules F on $\mathfrak{X}$ the base change theorem that

$$u^* R^i g_{\text{ét}*} t^* F \xrightarrow{\sim} R^i h_{\text{ét}*} v^* t^* F \qquad (7)$$

is an isomorphism of étale U-sheaves for every $i \geq 0$.

PROOF. Let $Y \to \mathfrak{Y}$ be a smooth presentation of $\mathfrak{Y}$. It suffices to prove that (7) is an isomorphism after making the base change to Y. Using notations as in the diagram

$$\begin{array}{ccccc} U' & \longrightarrow & V' & \longrightarrow & Y \\ \downarrow & \boxed{2} & \downarrow & \boxed{2} & \downarrow \\ U & \longrightarrow & V & \longrightarrow & \mathfrak{Y} \end{array}$$

this is easily proved by applying the smooth base change theorem for the four base changes $U' \to U$, $V' \to V$, $V' \to Y$ and $U' \to Y$. □

LEMMA 4.49. *Let $\mathfrak{S}$ be a site with topos of sheaves X. Let $\overline{X}$ be a full subcategory of $\mathfrak{S}$ such that $(X, \overline{X}, \pi)$ is a trivial c-topos. Let H be a presheaf on $\mathfrak{S}$ and $\widetilde{H}$ the associated sheaf. Then $\pi_* \widetilde{H}$ is the sheaf associated to $\pi_* H$.*

PROOF. This can be seen using the functor $\pi^! : \overline{X}^\wedge \to \mathfrak{S}^\wedge$, which is a right adjoint of $\pi_* : \mathfrak{S}^\wedge \to \overline{X}^\wedge$, and induces a functor $\pi^! : \overline{X} \to X$. (Apply Proposition 2.2 of [2, Exp. III] to $\overline{X} \to \mathfrak{S}$.) Here we denote by $\overline{X}^\wedge$ and $\mathfrak{S}^\wedge$ the categories of presheaves on $\overline{X}$ and $\mathfrak{S}$, respectively. □

PROPOSITION 4.50. *Let $f : \mathfrak{X} \to \mathfrak{Y}$ be a representable morphism of algebraic S-stacks. Let $\mathfrak{X} \to \mathfrak{X}'$ be a smooth representable epimorphism such that f factors through $\mathfrak{X}'$. Let $Y \to \mathfrak{Y}$ be a presentation of $\mathfrak{Y}$, denote by $f_Y : X \to Y$ the corresponding base change of f, and let F be an étale sheaf of A'-modules on $\mathfrak{X}$. Then for every $i \geq 0$ we have*

$$R^i f_{\mathfrak{X}'\text{-sm}*} F | Y_{\mathfrak{Y}\text{-sm}} = R^i f_{Y\,\text{ét}*} F_X$$

as sheaves on $Y_{\mathfrak{Y}\text{-sm}}$. Here $f_{\mathfrak{X}'\text{-sm}*}$ denotes the direct image functor

$$f_{\mathfrak{X}'\text{-sm}*} : \mathrm{Mod}(\mathfrak{X}_{\mathfrak{X}'\text{-sm}}, A') \longrightarrow \mathrm{Mod}(\mathfrak{Y}_{sm}, A').$$

In particular, $R^i f_{\mathfrak{X}'\text{-sm}*} F$ is an étale $\mathfrak{Y}$-sheaf. (In other words,

$$f : (\mathfrak{X}_{\mathfrak{X}'\text{-sm}}, \mathfrak{X}_{\text{ét}}) \longrightarrow (\mathfrak{Y}_{sm}, \mathfrak{Y}_{\text{ét}})$$

is a morphism of c-topoi of the second kind with respect to A'.)

PROOF. Consider the presheaf H on $\mathfrak{S}(\mathfrak{Y}_{\mathrm{sm}})$ given by

$$U \longmapsto H^i(f^*U_{\text{ét}}, F),$$

for $U \in \mathrm{ob}\,\mathfrak{S}(\mathfrak{Y}_{\mathrm{sm}})$. Compare H with the presheaf H' given by

$$U \longmapsto H^i(f^*U_{\mathfrak{X}'\text{-sm}}, F).$$

There is an obvious map $H \to H'$, which induces a bijection $H(U) \to H'(U)$ whenever $(f^*U_{\mathfrak{X}'\text{-sm}}, f^*U_{\text{ét}})$ is a trivial c-topos, so whenever U is a scheme. Since the schemes in $\mathfrak{S}(\mathfrak{Y}_{\mathrm{sm}})$ form a generating family, $H \to H'$ induces an isomorphism on the associated sheaves. This proves that the sheaf $\widetilde{H}$ associated to H is $R^i f_{\mathfrak{X}'\text{-sm}*} F$.

Now fix for the moment a $U \in \mathrm{ob}\,\mathfrak{S}(\mathfrak{Y}_{\mathrm{sm}})$, such that U is a scheme. Since $(U_{\mathfrak{Y}\text{-sm}}, U_{\text{ét}})$ is a trivial c-topos, we have by Lemma 4.49 that $\widetilde{H}$ restricted to $U_{\text{ét}}$ is the sheaf associated to the presheaf

$$U' \longmapsto H^i(f^*U'_{\text{ét}}, F),$$

which is none other than $R^i f_{U\,\text{ét}*} F$. Evaluating on global sections, we conclude that

$$\widetilde{H}(U) = R^i f_{U\,\text{ét}*} F(U).$$

Now assuming we are given a morphism $u : U \to Y$ in $\mathfrak{S}(\mathfrak{Y}_{\mathrm{sm}})$, then we have by Lemma 4.48 that

$$R^i f_{U\,\text{ét}*} F = u^* R^i f_{Y\,\text{ét}*} F,$$

so that

$$\widetilde{H}(U) = R^i f_{Y\,\text{ét}*} F(U).$$

This is what we needed to prove. □

COROLLARY 4.51. *In the situation of the Proposition we have commutative diagrams of functors*

$$\begin{array}{ccc} D^+_{\text{ét}}(\mathfrak{X}_{x'\text{-sm}}, A') & \stackrel{Rf_{x'\text{-sm}*}}{\longrightarrow} & D^+_{\text{ét}}(\mathfrak{Y}_{\text{sm}}, A') \\ \sigma_*\pi^* \downarrow & & \downarrow \sigma_*\pi^* \\ D^+_{\text{cart}}(X_{\bullet\,\text{ét}}, A') & \stackrel{Rf_{\bullet\,\text{ét}*}}{\longrightarrow} & D^+_{\text{cart}}(Y_{\bullet\,\text{ét}}, A') \end{array}$$

and

$$\begin{array}{ccc} D^+_{\text{ét}}(\mathfrak{X}^{\mathbb{N}}_{x'\text{-sm}}, \Lambda) & \stackrel{Rf^{\mathbb{N}}_{x'\text{-sm}*}}{\longrightarrow} & D^+_{\text{ét}}(\mathfrak{Y}^{\mathbb{N}}_{\text{sm}}, \Lambda) \\ \sigma_*\pi^* \downarrow & & \downarrow \sigma_*\pi^* \\ D^+_{\text{cart}}(X^{\mathbb{N}}_{\bullet\,\text{ét}}, \Lambda) & \stackrel{Rf^{\mathbb{N}}_{\bullet\,\text{ét}*}}{\longrightarrow} & D^+_{\text{cart}}(Y^{\mathbb{N}}_{\bullet\,\text{ét}}, \Lambda). \end{array}$$

Let us now fix a finite type algebraic S-stack $\mathfrak{X}$. Choosing a presentation X of $\mathfrak{X}$, recall that we have a d-structure on $|\mathfrak{X}|$ assigning to the locally closed substack $\mathfrak{V}$ of $\mathfrak{X}$ the t-category $D^+_{\text{cart}}(V_{\bullet\,\text{ét}}, A')$, where $V_\bullet$ is the simplicial algebraic space defined by the presentation V of $\mathfrak{V}$ induced by X. Considering this d-structure as a fibered category over $\text{lc}\,|\mathfrak{X}|_{\text{op}}$ using the direct image functors as in Remark 2.3, Corollary 4.51 shows that we may define a fibered category over $\text{lc}\,|\mathfrak{X}|_{\text{op}}$ whose fiber over $\mathfrak{V} \in \text{ob}\,\text{lc}\,|\mathfrak{X}|_{\text{op}}$ is $D^+_{\text{ét}}(\mathfrak{V}_{\text{sm}}, A')$ and whose pullback functors are the direct image functors. This is then obviously a d-structure on $|\mathfrak{X}|$.

DEFINITION 4.52. The d-structure thus constructed is called the *étale-smooth A'-d-structure on $\mathfrak{X}$*. We denote it by

$$\mathfrak{V} \longmapsto D^+_{\text{ét}}(\mathfrak{V}_{\text{sm}}, A')$$

or simply $D^+_{\text{ét}}(\mathfrak{X}_{\text{sm}}, A')$, by heavy abuse of notation.

Clearly, $D^+_{\text{lcc}}(\mathfrak{X}_{\text{sm}}, A') \subset D^+_{\text{ét}}(\mathfrak{X}_{\text{sm}}, A')$ defines a cd-structure on $|\mathfrak{X}|$. It is tractable as we saw in Section 4.

Let A be a discrete valuation ring whose residue characteristic is invertible on S, where S is of finite type over a regular base of dimension zero or one. In the same way as above, Corollary 4.51 gives rise to an A-d-structure on $|\mathfrak{X}|$, given by

$$\mathfrak{V} \longmapsto D^+_{\text{ét}}(\mathfrak{V}^{\mathbb{N}}_{\text{sm}}, \Lambda).$$

It passes to an A-d-structure

$$\mathfrak{V} \longmapsto \mathbb{D}^+_c(\mathfrak{V}_{\text{sm}}, A),$$

by the results of Section 4. (As usual, Λ denotes the projective system $\Lambda = (A/\ell^{n+1})_n$.)

DEFINITION 4.53. We call $\mathfrak{V} \longmapsto \mathbb{D}^+_c(\mathfrak{V}_{\text{sm}}, A)$ the *constructible ℓ-adic d-structure on $\mathfrak{X}$*.

6. The Étale-Smooth cd-Structure

Let us continue with consequences of Proposition 4.50.

COROLLARY 4.54. *Let $f : \mathfrak{X} \to \mathfrak{Y}$ be an arbitrary morphism of algebraic S-stacks. Then*

$$(f_{sm}, f_{\acute{e}t}) : (\mathfrak{X}_{sm}, \mathfrak{X}_{\acute{e}t}) \longrightarrow (\mathfrak{Y}_{sm}, \mathfrak{Y}_{\acute{e}t})$$

is a morphism of c-topoi of the second kind with respect to A'.

PROOF. Let F be an étale sheaf of A'-modules on $\mathfrak{X}$ and $X \to \mathfrak{X}$ a presentation of $\mathfrak{X}$. Let $X_\bullet$ be the corresponding simplicial algebraic space. Denote for every $p \in \mathbb{N}$ the induced morphism by $f_p : X_p \to \mathfrak{Y}$. Let $U \in \mathrm{ob}\,\mathfrak{S}(\mathfrak{Y}_{sm})$. Then we get an induced simplicial object $X_\bullet \times_\mathfrak{X} f^*U$ in $\mathfrak{S}(\mathfrak{X}_{sm})$ and whence a spectral sequence

$$E_1^{p,q} = H^q((X_p \times f^*U)_{\mathfrak{x}\text{-}sm}, F) \Longrightarrow H^{p+q}(f^*U_{\mathfrak{x}\text{-}sm}, F).$$

Note that $X_p \times f^*U = f_p^*U$, so that our spectral sequence may be written

$$E_1^{p,q} = H^q(f_p^*U_{\mathfrak{x}\text{-}sm}, F) \Longrightarrow H^{p+q}(f^*U_{\mathfrak{x}\text{-}sm}, F).$$

Letting U vary and passing to the associated sheaf on $\mathfrak{S}(\mathfrak{Y}_{sm})$, we get a spectral sequence

$$E_1^{p,q} = R^q f_{p\,\mathfrak{x}\text{-}sm*} F \Longrightarrow R^{p+q} f_{sm*} F.$$

We conclude using Proposition 4.50 and the fact that f_p is representable, for all $p \geq 0$. $\square$

COROLLARY 4.55. *In the situation of Proposition 4.50 we have a commutative diagram of functors*

$$\begin{array}{ccc} D_{\acute{e}t}^+(\mathfrak{X}_{\mathfrak{x}'\text{-}sm}, A') & \xrightarrow{Rf_{\mathfrak{x}'\text{-}sm*}} & D_{\acute{e}t}^+(\mathfrak{Y}_{sm}, A') \\ \pi_{X*}\rho^* \downarrow & & \downarrow \pi_{Y*}\tau^* \\ D^+(X_{\acute{e}t}, A') & \xrightarrow{Rf_{\acute{e}t*}} & D^+(Y_{\acute{e}t}, A'). \end{array}$$

Here ρ^* denotes the restriction of a sheaf from $\mathfrak{X}_{\mathfrak{x}'\text{-}sm}$ to $X_{\mathfrak{x}'\text{-}sm}$ and τ^* denotes the restriction of a sheaf from $\mathfrak{Y}_{sm}$ to $Y_{\mathfrak{Y}\text{-}sm}$. Finally, $\pi_X : X_{\mathfrak{x}'\text{-}sm} \to X_{\acute{e}t}$ and $\pi_Y : Y_{\mathfrak{Y}\text{-}sm} \to Y_{\acute{e}t}$ are the structure morphism of these two trivial c-topoi.

LEMMA 4.56. *Let $f : \mathfrak{X} \to \mathfrak{Y}$ be a representable morphism of algebraic S-stacks and $g : \mathfrak{Y} \to \mathfrak{Z}$ a further morphism of algebraic S-stacks. Let Y be presentation of $\mathfrak{Y}$ and X the induced presentation of $\mathfrak{X}$, giving rise to the following diagram of functors*

$$\begin{array}{ccccc} X_{\mathfrak{x}\text{-}sm} & \xrightarrow{f_{sm*}} & Y_{\mathfrak{Y}\text{-}sm} & & \\ a^* \uparrow & & \uparrow b^* & \searrow g_{sm*} & \\ \mathfrak{X}_{sm} & \xrightarrow{\widetilde{f}_{sm*}} & \mathfrak{Y}_{sm} & \xrightarrow{\widetilde{g}_{sm*}} & \mathfrak{Z}_{sm}. \end{array}$$

Then for every $M \in \mathrm{ob}\, D^+_{\text{ét}}(\mathfrak{X}_{sm}, A')$ the natural homomorphism

$$Rg_{sm*}Rf_{sm*}a^*M \longrightarrow R(g \circ f)_{sm*}a^*M$$

is an isomorphism. Here $(g \circ f)_{sm*}$ is the direct image functor from $X_{\mathfrak{x}\text{-}sm}$ to $\mathfrak{Z}_{sm}$.

PROOF. Letting $u : Z \to \mathfrak{Z}$ be a presentation of $\mathfrak{Z}$, it suffices to prove that

$$u^*Rg_{\mathrm{sm}*}Rf_{\mathrm{sm}*}a^*M \longrightarrow u^*R(g \circ f)_{\mathrm{sm}*}a^*M$$

is an isomorphism. This may be proved by applying Corollary 4.55 to the representable morphisms $Y \to \mathfrak{Z}$, $X \to \mathfrak{Z}$ and $\mathfrak{X} \to \mathfrak{Y}$, using notations as in

$$\begin{array}{ccccc} X' & \longrightarrow & Y' & & \\ \downarrow & \square & \downarrow & \searrow & \\ X & \longrightarrow & Y & \square & Z \\ \downarrow & \square & \downarrow & \searrow & \downarrow \\ \mathfrak{X} & \longrightarrow & \mathfrak{Y} & \longrightarrow & \mathfrak{Z}. \end{array}$$

One reduces to the composition $X'_{\text{ét}} \to Y'_{\text{ét}} \to Z_{\text{ét}}$ of honest morphisms of topoi. $\square$

PROPOSITION 4.57. *Let $f : \mathfrak{X} \to \mathfrak{Y}$ be a morphism of algebraic S-stacks. Then for every $g : \mathfrak{Y} \to \mathfrak{Z}$ and any $M \in \mathrm{ob}\, D^+_{\text{ét}}(\mathfrak{X}_{sm}, A')$ the canonical morphism*

$$Rg_{sm*}Rf_{sm*}M \longrightarrow R(g \circ f)_{sm*}M$$

is an isomorphism. Here we endow all three stacks $\mathfrak{X}$, $\mathfrak{Y}$ and $\mathfrak{Z}$ with their absolute smooth topoi.

PROOF. We will use the first devissage lemma (Proposition 4.16). Thus we have to check four facts.

Let us first prove that the proposition holds when f is smooth. In this case $f_{\mathrm{sm}*}$ has an exact left adjoint, so it takes injective sheaves of A'-modules to injective sheaves of A'-modules. So the proposition holds even for $M \in \mathrm{ob}\, D^+(\mathfrak{X}_{\mathrm{sm}}, A')$.

Secondly, let us assume that f is representable. Let Y be a presentation of $\mathfrak{Y}$ and X the induced presentation of $\mathfrak{X}$. Let $X_\bullet$ and $Y_\bullet$ be the corresponding simplicial algebraic spaces and use notations as in the diagram

$$\begin{array}{ccccc} X_p & \xrightarrow{f_p} & Y_p & & \\ \downarrow & & \downarrow & \searrow^{g_p} & \\ \mathfrak{X} & \xrightarrow{f} & \mathfrak{Y} & \xrightarrow{g} & \mathfrak{Z}, \end{array}$$

with $h_p = g_p \circ f_p$ and $h = g \circ f$. For every $M \in \mathrm{ob}\, D^+(\mathfrak{Y}_{\mathrm{sm}}, A')$ we have a spectral sequence

$$E_1^{p,q} = h^q Rg_{p\,\mathfrak{Y}\text{-sm}*}(M|Y_{p\,\mathfrak{Y}\text{-sm}}) \Longrightarrow h^{p+q} Rg_{\mathrm{sm}*}M. \qquad (8)$$

Letting $M \in \mathrm{ob}\, D^+(\mathfrak{X}_{\mathrm{sm}}, A')$, and applying (8) to $Rf_{\mathrm{sm}*}M$, we get a spectral sequence in $\mathrm{Mod}(\mathfrak{Z}_{\mathrm{sm}}, A')$

$$E_1^{p,q} = h^q Rg_{p_{\mathfrak{Y}\text{-sm}*}}(Rf_{\mathrm{sm}*}M|Y_{p_{\mathfrak{Y}\text{-sm}}}) \Longrightarrow h^{p+q} Rg_{\mathrm{sm}*}Rf_{\mathrm{sm}*}M.$$

Now by Lemma 4.56 we have

$$Rg_{p_{\mathfrak{Y}\text{-sm}*}}(Rf_{\mathrm{sm}*}M|Y_{p_{\mathfrak{Y}\text{-sm}}}) = Rh_{p_{\mathfrak{X}\text{-sm}*}}(M|X_{p_{\mathfrak{X}\text{-sm}}}),$$

so that our spectral sequence reads

$$E_1^{p,q} = h^q Rh_{p_{\mathfrak{X}\text{-sm}*}}(M|X_{p_{\mathfrak{X}\text{-sm}}}) \Longrightarrow h^{p+q} Rg_{\mathrm{sm}*}Rf_{\mathrm{sm}*}M.$$

But this sequence abuts to $h^{p+q} Rh_{\mathrm{sm}*}M$ which proves our claim.

As third part let us consider a diagram

$$\begin{array}{ccccc}
& \mathfrak{U} & & & \\
& {\scriptstyle j}\downarrow\;\;\searrow{\scriptstyle g} & & & \\
& \mathfrak{X} & \xrightarrow{f} & \mathfrak{Y} & \xrightarrow{h'} & \mathfrak{Y}' \\
& {\scriptstyle i}\uparrow\;\;\nearrow{\scriptstyle h} & & & \\
& \mathfrak{Z} & & &
\end{array}$$

and assume that the proposition holds for g and h. To prove it for f, let $M \in \mathrm{ob}\, D^+_{\mathrm{\acute{e}t}}(\mathfrak{X}_{\mathrm{sm}}, A')$. We have a distinguished triangle in $D^+_{\mathrm{\acute{e}t}}(\mathfrak{X}_{\mathrm{sm}}, A')$

$$\begin{array}{ccc}
& Rj_*j^*M & \\
& \nearrow \quad \searrow & \\
i_*Ri^!M & \longrightarrow & M,
\end{array}$$

using the étale-smooth A'-d-structure on $\mathfrak{X}$ (see Definition 4.52). Since we already proved the proposition for immersions, we derive a distinguished triangle in $D^+_{\mathrm{\acute{e}t}}(\mathfrak{Y}_{\mathrm{sm}}, A')$

$$\begin{array}{ccc}
& Rg_*j^*M & \\
& \nearrow \quad \searrow & \\
Rh_*Ri^!M & \longrightarrow & Rf_*M,
\end{array}$$

and one in $D^+_{\mathrm{\acute{e}t}}(\mathfrak{Y}'_{\mathrm{sm}}, A')$

$$\begin{array}{ccc}
& R(h'g)_*j^*M & \\
& \nearrow \quad \searrow & \\
R(h'h)_*Ri^!M & \longrightarrow & R(h'f)_*M.
\end{array}$$

Applying Rh'_* to the first one and comparing with the second one we get the result.

Finally, it is a pure formality that the proposition holds for $f' \circ f : \mathfrak{X} \to \mathfrak{X}''$ if it holds for $f : \mathfrak{X} \to \mathfrak{X}'$ and $f' : \mathfrak{X}' \to \mathfrak{X}''$. $\square$

COROLLARY 4.58. *Let* $f : \mathfrak{X} \to \mathfrak{Y}$ *and* $g : \mathfrak{Y} \to \mathfrak{Z}$ *be morphisms of algebraic S-stacks. Then for every* $M \in \mathrm{ob}\, D^+_{\mathrm{\acute{e}t}}(\mathfrak{X}^{\mathbb{N}}_{sm}, \Lambda)$ *the canonical morphism*

$$Rg^{\mathbb{N}}_{sm*} Rf^{\mathbb{N}}_{sm*} M \longrightarrow R(gf)^{\mathbb{N}}_{sm*} M$$

in $D^+_{\mathrm{\acute{e}t}}(\mathfrak{Z}^{\mathbb{N}}_{sm}, \Lambda)$ *is an isomorphism.*

Let $f : \mathfrak{X} \to \mathfrak{Y}$ b a morphism of algebraic S-stacks. If we have a commutative diagram

$$\begin{array}{ccc} X & \xrightarrow{f_0} & Y \\ \downarrow & & \downarrow \\ \mathfrak{X} & \xrightarrow{f} & \mathfrak{Y}, \end{array} \qquad (9)$$

where $X \to \mathfrak{X}$ and $Y \to \mathfrak{Y}$ are presentations, then the isomorphism of Remark 4.46 allows us to define a functor

$$f^* : D^+_{\text{ét}}(\mathfrak{Y}_{\text{sm}}, A') \to D^+_{\text{ét}}(\mathfrak{X}_{\text{sm}}, A'),$$

which arises via transport of structure from

$$f^*_{\bullet\,\text{ét}} : D^+_{\text{cart}}(Y_{\bullet\,\text{ét}}, A') \to D^+_{\text{cart}}(X_{\bullet\,\text{ét}}, A').$$

Similarly, the diagram (9) defines a functor

$$f^{\mathbb{N}*} : D^+_{\text{ét}}(\mathfrak{Y}^{\mathbb{N}}_{\text{sm}}, \Lambda) \to D^+_{\text{ét}}(\mathfrak{X}^{\mathbb{N}}_{\text{sm}}, \Lambda).$$

PROPOSITION 4.59. *In this way every diagram (9) gives rise to a left adjoint of*

$$Rf_* : D^+_{\text{ét}}(\mathfrak{X}_{sm}, A') \to D^+_{\text{ét}}(\mathfrak{Y}_{sm}, A')$$

and of

$$Rf^{\mathbb{N}}_* : D^+_{\text{ét}}(\mathfrak{X}^{\mathbb{N}}_{sm}, \Lambda) \to D^+_{\text{ét}}(\mathfrak{Y}^{\mathbb{N}}_{sm}, \Lambda).$$

PROOF. If f is smooth, then $f_{\text{sm}} : \mathfrak{X}_{\text{sm}} \to \mathfrak{Y}_{\text{sm}}$ and $f^{\mathbb{N}}_{\text{sm}} : \mathfrak{X}^{\mathbb{N}}_{\text{sm}} \to \mathfrak{Y}^{\mathbb{N}}_{\text{sm}}$ are morphism of c-topoi of the first kind. Thus f^*_{sm} and $f^{\mathbb{N}\,*}_{\text{sm}}$ give rise to left adjoints as required. That they are compatible with the pullback functors arising from diagram (9) is easily proved using similar arguments as in Proposition 3.65.

If f is representable, then $f_{\bullet\,\text{ét}} : \text{top}\,X_{\bullet\,\text{ét}} \to \text{top}\,Y_{\bullet\,\text{ét}}$ and $f^{\mathbb{N}}_{\bullet\,\text{ét}} : \text{top}(X_{\bullet\,\text{ét}})^{\mathbb{N}} \to \text{top}(Y_{\bullet\,\text{ét}})^{\mathbb{N}}$ are morphisms of c-topoi of the first and second kind. Thus $f_{\bullet\,\text{ét}*}$ and $f^{\mathbb{N}}_{\bullet\,\text{ét}*}$ are right adjoints of our pullback functors and so the proposition follows from Corollary 4.51.

Let $f : \mathfrak{X} \to \mathfrak{Y}$ and $g : \mathfrak{Y} \to \mathfrak{Z}$ be morphisms of algebraic S-stacks and assume that the Proposition holds for f and g. That the proposition then holds also for $g \circ f$ follows easily from Proposition 4.57 and its Corollary 4.58.

Now let $\mathfrak{X}$ be the disjoint union of a closed substack $i : \mathfrak{Z} \to \mathfrak{X}$ and its open complement $j : \mathfrak{U} \to \mathfrak{X}$. Let us assume that the proposition holds for $f \circ i$ and $f \circ j$. Let $M \in \text{ob}\, D^+_{\text{ét}}(\mathfrak{X}_{\text{sm}}, A')$. We get distinguished triangles in $D^+_{\text{ét}}(\mathfrak{X}_{\text{sm}}, A')$ and $D^+_{\text{ét}}(\mathfrak{Y}_{\text{sm}}, A')$

$$\begin{array}{ccc} & Rj_*j^*M & \\ & \nearrow \quad \nwarrow & \\ i_*Ri^!M & \longrightarrow & M, \end{array}$$

$$\begin{array}{ccc} & R(fj)_*j^*M & \\ & \nearrow \quad \nwarrow & \\ R(fi)_*Ri^!M & \longrightarrow & Rf_*M. \end{array}$$

Apply $\text{Hom}(f^*N, \cdot)$ to the first and $\text{Hom}(N, \cdot)$ to the second to get the required result.

Finally, the proposition now follows from the first devissage lemma Proposition 4.16. □

So from now on we denote by f^* any left adjoint of $Rf_* : D^+_{\text{ét}}(\mathfrak{X}_{\text{sm}}, A') \to D^+_{\text{ét}}(\mathfrak{Y}_{\text{sm}}, A')$.

COROLLARY 4.60 (Smooth base change). *Consider a 2-cartesian diagram of S-stacks*

$$\begin{array}{ccc} \mathfrak{X}' & \xrightarrow{f'} & \mathfrak{Y}' \\ v \downarrow & & \downarrow u \\ \mathfrak{X} & \xrightarrow{f} & \mathfrak{Y}, \end{array}$$

where $\mathfrak{Y}' \to \mathfrak{Y}$ is smooth. Then for every $M \in \text{ob}\, D^+_{\text{ét}}(\mathfrak{X}_{sm}, A')$ we have a canonical isomorphism

$$u^* Rf_* M \xrightarrow{\sim} Rf'_* v^* M.$$

A similar statement holds for objects of $D^+_{\text{ét}}(\mathfrak{X}^{\mathbb{N}}_{sm}, \Lambda)$.

PROOF. First of all, the base change homomorphism exists since we have all the adjoint and composition properties needed for its definition, by Propositions 4.57 and 4.59. To prove that it is an isomorphism it suffices to consider the case that $M = F$ is an étale sheaf of A' modules on $\mathfrak{X}$. Clearly, we may also assume $u : \mathfrak{Y}' \to \mathfrak{Y}$ to be representable.

Let us first consider the case that f is representable. Choose a cartesian diagram of presentations

$$\begin{array}{ccc} Y' & \xrightarrow{u'} & Y \\ \downarrow & & \downarrow \\ \mathfrak{Y}' & \xrightarrow{u} & \mathfrak{Y}, \end{array}$$

and let

$$\begin{array}{ccc} X' & \xrightarrow{v'} & X \\ \downarrow & & \downarrow \\ \mathfrak{X}' & \xrightarrow{v} & \mathfrak{X} \end{array}$$

be the pullback via f. It suffices to prove that we have an isomorphism after pullback to Y'. This is easily reduced to the usual smooth base change theorem by applying Proposition 4.50 to the morphism $f : \mathfrak{X} \to \mathfrak{Y}$ and the presentations Y and Y' of $\mathfrak{Y}$, respectively.

Now the general case can be proved using the spectral sequence

$$E^{p,q}_1 = R^q f_{p*} F \Longrightarrow R^{p+q} f_* F,$$

associated to a presentation $X \to \mathfrak{X}$, where f_p is the induced map $X_p \to \mathfrak{Y}$. □

7. The ℓ-Adic d-Structure

Let $f : \mathfrak{X} \to \mathfrak{Y}$ be a morphism of algebraic S-stacks of finite type. From Propositions 4.57 and 4.59 it is now clear that f induces a morphism of the étale-smooth A'-cd-structures on $\mathfrak{X}$ and $\mathfrak{Y}$. We also get a morphism from the d-structure $\mathfrak{V} \mapsto D_{\text{ét}}^+(\mathfrak{V}_{\text{sm}}^{\mathbb{N}}, \Lambda)$ on $\mathfrak{X}$ to the d-structure $\mathfrak{W} \mapsto D_{\text{ét}}^+(\mathfrak{W}_{\text{sm}}^{\mathbb{N}}, \Lambda)$ on $\mathfrak{Y}$. Our goal is to prove that it passes to the constructible ℓ-adic d-structures. For this it will be essential that the morphism of the étale-smooth Λ_0-cd-structures is tractable. To prove this (Theorem 4.70) we need some preliminaries.

If not indicated otherwise we think of algebraic S-stacks as endowed with their étale-smooth A'-cd-structure. For a morphism of algebraic stacks we will denote the induced morphism of étale-smooth cd-structures by the same letter.

7.1. Closed L-Stratifications.
We will need L-stratifications satisfying an additional hypothesis.

DEFINITION 4.61. Let $\mathfrak{X}$ be an algebraic S-stack of finite type and A a ring. We call a family $\mathcal{L}$ of simple lcc sheaves of A-modules on $\mathfrak{X}_{\text{ét}}$ *closed*, if for every finite family $L_1, \ldots L_n$ of elements of $\mathcal{L}$ the tensor product $L_1 \otimes \ldots \otimes L_n$ has Jordan-Hölder components isomorphic to elements of $\mathcal{L}$.

Clearly, intersections of closed families of simple lcc sheaves of A-modules are closed.

LEMMA 4.62. *Let $f : \mathfrak{Y} \to \mathfrak{X}$ be a finite étale morphism of algebraic S-stacks of finite type. Then the set of isomorphism classes of simple lcc sheaves of A-modules on $\mathfrak{X}_{\text{ét}}$ that are trivialized by $\mathfrak{Y}$ is finite.*

PROOF. We may assume that f is a principal bundle with finite structure group G. Let k be the residue field of A. Any simple object of $\text{Mod}_{\text{lcc}}(\mathfrak{X}_{\text{ét}}, A)$ is in fact a sheaf of k-vector spaces. So the simple lcc modules trivialized by $\mathfrak{Y}$ are equivalent to simple finite left $k[G]$-modules. Every simple module over $k[G]$ is cyclic and hence isomorphic to $k[G]/\mathfrak{m}$, where $\mathfrak{m}$ is a maximal left ideal of $k[G]$. But these are finite in number. $\square$

COROLLARY 4.63. *Let $\mathcal{L}$ be a finite family of simple lcc sheaves of A-modules on $\mathfrak{X}_{\text{ét}}$. Then there exists a finite such family $\widetilde{\mathcal{L}}$ containing $\mathcal{L}$, which is closed. The smallest such $\widetilde{\mathcal{L}}$ is called the* closed hull *of $\mathcal{L}$.*

PROOF. Let $\mathfrak{Y}$ be a finite étale cover of $\mathfrak{X}$, trivializing all elements of $\mathcal{L}$. Then let $\widetilde{\mathcal{L}}$ represent the set of all simple lcc sheaves on $\mathfrak{X}$, trivialized by $\mathfrak{Y}$. $\square$

DEFINITION 4.64. A pre-L-stratification $(\mathcal{S}, \mathcal{L})$ of $\mathfrak{X}_{\text{ét}}$ is called *closed*, if for every $\mathfrak{V} \in \mathcal{S}$ the set of lcc sheaves $\mathcal{L}(\mathfrak{V})$ is closed on $\mathfrak{V}_{\text{ét}}$.

LEMMA 4.65. *Every L-stratification of $\mathfrak{X}_{\text{ét}}$ admits a refinement which is a closed L-stratification.*

PROOF. Going through the proof of Lemma 2.10, we notice that after we refine $(\mathcal{S}, \mathcal{L})$ to an L-stratification over $U \cup V$ (in the notation of loc. cit.) we may pass to the closed hull of $\mathcal{L}(V)$ and thus ensure that all $\mathcal{L}(V)$ become closed. □

Consider the following setup. Let S be a scheme of finite type over a regular scheme of dimension zero or one, and k a finite field, whose characteristic is invertible on S. Let $\pi : G \to \mathfrak{X}$ be a smooth group $\mathfrak{X}$-space with connected fibers, where $\mathfrak{X}$ is an algebraic S-stack of finite type. Assume that the relative dimension d of G over $\mathfrak{X}$ is constant.

We know that $R^p\pi_*k$ is a constructible sheaf of k-vector spaces on $\mathfrak{X}_{\text{ét}}$, for $q = 0, \ldots, 2d$ and zero for $q > 2d$ (see for example Corollary 4.60). Let us in fact assume that $R^q\pi_*k$ is lcc, for every $q \geq 0$. Let $\mathcal{L}_0$ be a finite set of simple lcc sheaves of k-vector spaces on $\mathfrak{X}_{\text{ét}}$ such that the $R^q\pi_*k$ are trivialized by $\mathcal{L}$, meaning that the Jordan-Hölder components of $R^q\pi_*k$ are isomorphic to elements of $\mathcal{L}$, for all $q \geq 0$. Now let F be an lcc sheaf of k-vector spaces on $\mathfrak{X}_{\text{ét}}$, and $\mathcal{L}$ a finite set of simple lcc sheaves of k-vector spaces on $\mathfrak{X}_{\text{ét}}$ such that F is trivialized by $\mathcal{L}$. Assume that $\mathcal{L}_0 \subset \mathcal{L}$ and that $\mathcal{L}$ is closed.

We introduce the following notation. Let $G_0 = \mathfrak{X}$ and $G_n = G_{n-1} \times_{\mathfrak{X}} G$, for $n \geq 1$, and let $\pi_n : G_n \to \mathfrak{X}$ be the structure morphism. The pullback of F via π_n to G_n will be denoted by F_n.

LEMMA 4.66. *For every $n \geq 0$ the sheaf of k-vector spaces $R^q\pi_{n*}(F_n)$ is lcc and trivialized by $\mathcal{L}$, for every $q \geq 0$.*

PROOF. We will proceed by induction on n, the case $n = 0$ being trivial. For $n \geq 1$ consider the following cartesian diagram of algebraic S-stacks

$$\begin{array}{ccc} G_n & \xrightarrow{p_2} & G_{n-1} \\ p_1 \downarrow & & \downarrow \pi_{n-1} \\ G & \xrightarrow{\pi} & \mathfrak{X}. \end{array}$$

By the smooth base change theorem (Corollary 4.60) we have

$$R^q p_{1*} F_n = \pi^* R^q \pi_{n-1*}(F_{n-1}),$$

for every $q \geq 0$. So the E_2-term of the Leray spectral sequence of the composition $\pi_n = \pi \circ p_1$ is given by

$$\begin{aligned} E_2^{p,q} &= R^p\pi_* R^q p_{1*} F_n \\ &= R^p\pi_* \pi^* R^q \pi_{n-1*}(F_{n-1}) \\ &= R^p\pi_* k \otimes_k R^q \pi_{n-1*}(F_{n-1}) \end{aligned}$$

by the projection formula, which we may apply since $R^q\pi_{n-1*}(F_{n-1})$ is locally free, by the induction hypothesis. Since we have assumed $R^p\pi_*k$ to be lcc, trivialized by $\mathcal{L}$, the induction hypothesis implies that $E_2^{p,q}$ is also trivialized by $\mathcal{L}$, for all p, q. Passing through the spectral sequence the claim follows. □

Now consider the classifying stack $BG = B(G/\mathfrak{X})$ of G. Denote the structure morphism by $f : BG \to \mathfrak{X}$.

COROLLARY 4.67. *Let $\mathcal{L}$ be as above. Let $M \in \mathrm{ob}\, D^+_{\mathrm{lcc}}(BG_{sm}, k)$ be trivialized by $\mathcal{L}$. Then Rf_*M is an object of $D^+_{\mathrm{lcc}}(\mathfrak{X}_{sm}, k)$, trivialized by $\mathcal{L}$.*

PROOF. First note that without loss of generality we may assume $M = F$ to be an étale sheaf on BG, which comes from $\mathfrak{X}$ by Proposition 4.29. Then we need only apply the spectral sequence

$$E_1^{p,q} = R^q \pi_{p*} F_p \Longrightarrow R^{p+q} f_{sm*}(f^*F),$$

which is just the spectral sequence of Čech cohomology associated to the one-object covering $\mathfrak{X} \to BG$ of BG, given by the universal G-torsor. □

7.2. Some more auxiliary results.

LEMMA 4.68. *Let $f : \mathfrak{X} \to \mathfrak{Y}$ be a universal homeomorphism. Then the functor*

$$Rf_* : D^+_{\acute{e}t}(\mathfrak{X}_{sm}, A') \longrightarrow D^+_{\acute{e}t}(\mathfrak{Y}_{sm}, A')$$

is an equivalence of categories with quasi-inverse f^*.

PROOF. By Remark 4.26 this reduces to proving that $R^i f_* F = 0$, for $i > 0$ and F an étale sheaf on $\mathfrak{X}$. This, on the other hand, reduces by Proposition 4.57 to the representable case. But then by Corollary 4.60 we may base change to a presentation of $\mathfrak{Y}$ and use the étale topology to compute $R^i f_* F$. Then we are done since f is finite. □

LEMMA 4.69. *Let $f : \mathfrak{X} \to \mathfrak{Y}$ be gerbe-like such that Δ_f has components and A any ring. Then for any pre-L-stratification $(\mathcal{S}, \mathcal{L})$ of $\mathfrak{X}$ there exists a pre-L-stratification $(\mathcal{S}', \mathcal{L}')$ of $\mathfrak{Y}$, such that for a sheaf of A-modules F on $\mathfrak{Y}_{\acute{e}t}$ we have that f^*F being $(\mathcal{S}, \mathcal{L})$-constructible implies that F is $(\mathcal{S}', \mathcal{L}')$-constructible.*

PROOF. Choose a maximal element V of $\mathcal{S}$. Then $V \to \mathfrak{X}$ is an open immersion and $V \to \mathfrak{Y}$ is smooth. Let $V' \subset \mathfrak{Y}$ be the image of V in $\mathfrak{Y}$, which is an open substack of $\mathfrak{Y}$, and let $\mathfrak{Y}'$ be a closed complement of V' in $\mathfrak{Y}$. Then is suffices to prove the lemma for $V \to V'$ and the pullback of f via $\mathfrak{Y}' \to \mathfrak{Y}$. So by noetherian induction we reduce to the case that $\mathcal{S}$ contains only one stratum.

Let $\mathcal{L}(\mathfrak{X}) = \{L_1, \ldots, L_n\}$. Let $L_1, \ldots, L_s$ be those elements of $\mathcal{L}(\mathfrak{X})$ that come via f^* from $\mathfrak{Y}$. Choose $M_1, \ldots, M_s$ such that $f^*M_i = L_i$, for $i = 1, \ldots, s$. Then set $\mathcal{S}' = \{\mathfrak{Y}\}$ and $\mathcal{L}'(\mathfrak{Y}) = \{M_1, \ldots, M_s\}$. Now the claim follows from the fact that a simple lcc sheaf of A-modules on $\mathfrak{Y}_{\acute{e}t}$ remains simple after pullback to $\mathfrak{X}$. This follows from the following two facts.

(1) $f^* : \mathrm{Mod}(\mathfrak{Y}_{\acute{e}t}, A) \to \mathrm{Mod}(\mathfrak{X}_{\acute{e}t}, A)$ is fully faithful.
(2) For every $F \in \mathrm{ob}\,\mathrm{Mod}(\mathfrak{X}_{\acute{e}t}, A)$ the adjunction homomorphism $f^*f_*F \to F$ is a monomorphism.

Fact (1) follows easily using descent theory from the fact that $\Delta_f : \mathfrak{X} \to \mathfrak{X}_\mathfrak{Y}\mathfrak{X}$ is a flat epimorphism. To prove (2) we may factor $\mathfrak{X} \to \mathfrak{Y}$ as $\mathfrak{X} \to \mathfrak{X}' \to \mathfrak{Y}$, where $\mathfrak{X} \to \mathfrak{X}'$ is gerbe-like with connected fibers of the diagonal and $\mathfrak{X}' \to \mathfrak{Y}$ is étale (see Proposition 4.11). Since $\mathfrak{X} \to \mathfrak{X}'$ induces an isomorphism of étale topoi (Proposition 4.29), we are reduced to the étale case. Then it is not very hard to prove (2) using (1). □

7.3. The Main Theorem. Let k be a finite field whose characteristic is invertible on S, which is a scheme of finite type over some noetherian regular base of dimension zero or one.

THEOREM 4.70. *For every morphism of finite type algebraic S-stacks the induced morphism of étale-smooth k-cd-structures is tractable. (See Definitions 4.52 and 2.14.)*

PROOF. We will use the second devissage lemma (Proposition 4.17). By Lemma 4.69 condition (1) is satisfied by our property. Conditions (2) and (3) are satisfied by general theory of tractable morphisms of cd-structures, in particular Lemma 2.15. The theorem holds for representable morphisms by Proposition 4.44.

So let us assume that we are in case (5) and use notation as in Lemma 4.66 and Corollary 4.67. Note that by Proposition 4.29 L-stratifications of $\mathfrak{X}$ and BG coincide. Let $(\mathcal{S}, \mathcal{L})$ be a closed L-stratification of $\mathfrak{X}$ such that for every stratum $\mathfrak{V} \in \mathcal{S}$ we have that $R^i \pi_{\mathfrak{V}*} k$ is lcc, trivialized by $\mathcal{L}(\mathfrak{V})$, for all i. We claim that if $M \in \mathrm{ob}\, D^+_{(\mathcal{S},\mathcal{L})}(BG_\mathrm{sm}, k)$, then $Rf_* M$ is $(\mathcal{S}, \mathcal{L})$-constructible. This reduces to the case $M \in \mathrm{ob}\, \mathrm{Mod}_{(\mathcal{S},\mathcal{L})}(BG_\mathrm{\acute{e}t}, k)$, so that we may assume that $M = f^* F$, for an $(\mathcal{S}, \mathcal{L})$-constructible sheaf of k-vector spaces on $\mathfrak{X}_\mathrm{\acute{e}t}$. To prove that $Rf_* f^* F$ is $(\mathcal{S}, \mathcal{L})$-constructible, we may pass to the strata of $\mathcal{S}$ (using Lemma 2.12) and thus assume that $\mathfrak{X}$ contains only one stratum. Then we are in the case of Corollary 4.67. □

Now let A be a discrete valuation ring with residue field k and parameter ℓ. Note that by Nakayama's lemma every simple lcc sheaf of A-modules is a sheaf of k-vector spaces. Thus L-stratifications with respect to A and with respect to k coincide.

PROPOSITION 4.71. *Let $f : \mathfrak{X} \to \mathfrak{Y}$ be a morphism of finite type algebraic S-stacks. Let $(\mathcal{S}, \mathcal{L})$ be an L-stratification of $\mathfrak{X}_{\acute{e}t}$ and $(\mathcal{S}', \mathcal{L}')$ an L-stratification of $\mathfrak{Y}_{\acute{e}t}$ such that under*

$$Rf_* : D^+_{\acute{e}t}(\mathfrak{X}_{sm}, k) \longrightarrow D^+_{\acute{e}t}(\mathfrak{Y}_{sm}, k)$$

the $(\mathcal{S}, \mathcal{L})$-constructible objects map to $(\mathcal{S}', \mathcal{L}')$-constructible objects. (Existence of $(\mathcal{S}', \mathcal{L}')$ is guaranteed by Theorem 4.70 and Lemma 2.16.) Then the functor $Rf_* : D^+(\mathfrak{X}_{sm}, A) \to D^+(\mathfrak{Y}_{sm}, A)$ maps $(\mathcal{S}, \mathcal{L})$-constructible objects to $(\mathcal{S}', \mathcal{L}')$-constructible objects.

PROOF. Let $M \in \mathrm{ob}\, D^+_{(\mathcal{S},\mathcal{L})}(\mathfrak{X}_\mathrm{sm}, A)$. To prove that $Rf_* M \in \mathrm{ob}\, D^+_{(\mathcal{S}',\mathcal{L}')}(\mathfrak{Y}_\mathrm{sm}, A)$, we may assume that $M = F$ is an étale sheaf on $\mathfrak{X}$.

Induction on the length of the ℓ-adic filtration of F reduces to the case that F is a sheaf of k-vector spaces. Then we are done, since the derived functors of f_* commute with restriction of scalars as was noted in Remark 3.32. □

COROLLARY 4.72. *Let $f : \mathfrak{X} \to \mathfrak{Y}$ be a morphism of finite type algebraic S-stacks. Then $f : (\mathfrak{X}_{sm}, \mathfrak{X}_{\acute{e}t}) \longrightarrow (\mathfrak{Y}_{sm}, \mathfrak{Y}_{\acute{e}t})$ is a morphism of c-topoi of the third kind with respect to A.*

By this corollary the functor
$$Rf_*^{\mathbb{N}} : D_{\acute{e}t}^+(\mathfrak{X}_{sm}^{\mathbb{N}}, \Lambda) \to D_{\acute{e}t}^+(\mathfrak{Y}_{sm}^{\mathbb{N}}, \Lambda)$$
passes as in Remark 3.49 to an ℓ-adic derived functor
$$\mathbb{R} f_* : \mathbb{D}_c^+(\mathfrak{X}_{sm}, A) \longrightarrow \mathbb{D}_c^+(\mathfrak{Y}_{sm}, A)$$
between the categories of constructible A-complexes on $\mathfrak{X}_{sm}$ and $\mathfrak{Y}_{sm}$, respectively.

PROPOSITION 4.73. *Let*
$$f^{\mathbb{N}*} : D_{\acute{e}t}^+(\mathfrak{Y}_{sm}^{\mathbb{N}}, \Lambda) \to D_{\acute{e}t}^+(\mathfrak{X}_{sm}^{\mathbb{N}}, \Lambda)$$
be a left adjoint of
$$Rf^{\mathbb{N}}_* : D_{\acute{e}t}^+(\mathfrak{X}_{sm}^{\mathbb{N}}, \Lambda) \to D_{\acute{e}t}^+(\mathfrak{Y}_{sm}^{\mathbb{N}}, \Lambda)$$
constructed as in Proposition 4.59. Then the induced functor
$$f^* : \mathbb{D}_c^+(\mathfrak{Y}_{sm}, A) \longrightarrow \mathbb{D}_c^+(\mathfrak{X}_{sm}, A)$$
is a left adjoint of
$$\mathbb{R} f_* : \mathbb{D}_c^+(\mathfrak{X}_{sm}, A) \longrightarrow \mathbb{D}_c^+(\mathfrak{Y}_{sm}, A).$$

PROOF. This is a straightforward, if tedious, check. (Compare Remark 1.5.) □

COROLLARY 4.74 (Smooth base change). *Let $f' : \mathfrak{X}' \to \mathfrak{Y}'$ be induced by a smooth base change $u : \mathfrak{Y}' \to \mathfrak{Y}$. Then we have natural isomorphisms*
$$u^* \mathbb{R} f_* M \xrightarrow{\sim} \mathbb{R} f'_* v^* M,$$
for every $M \in \mathrm{ob}\,\mathbb{D}_c^+(\mathfrak{X}_{sm}, A)$. Here v is the morphism $v : \mathfrak{X}' \to \mathfrak{X}$.

PROOF. This follows immediately from Corollary 4.60. □

REMARK 4.75. It is now clear that a morphism of finite type algebraic S-stacks $f : \mathfrak{X} \to \mathfrak{Y}$ induces a morphism of the constructible ℓ-adic d-structures on $\mathfrak{X}$ and $\mathfrak{Y}$, respectively (see Definition 4.53). So we have finally achieved our main goal of constructing an ℓ-adic formalism for algebraic stacks. Additional properties enjoyed by our ℓ-adic formalism are smooth base change (Corollary 4.60) and the fact that a universal homeomorphism $f : \mathfrak{X} \to \mathfrak{Y}$ induces an equivalence of categories
$$\mathbb{R} f_* : \mathbb{D}_c^+(\mathfrak{X}_{sm}, A) \longrightarrow \mathbb{D}_c^+(\mathfrak{Y}_{sm}, A),$$

with quasi-inverse f^*. Moreover, if $\mathfrak{X}$ is a Deligne-Mumford stack, then
$$\mathbb{D}_c^+(\mathfrak{X}_{\mathrm{sm}}, A) = \mathbb{D}_c^+(\mathfrak{X}_{\mathrm{ét}}, A).$$

REMARK 4.76. Another consequence of Corollary 4.72 is that we have a commutative diagram
$$\begin{array}{ccc} D_c^+(\mathfrak{X}_{\mathrm{sm}}, A) & \longrightarrow & D_c^+(\mathfrak{Y}_{\mathrm{sm}}, A) \\ \sigma_*\pi^* \downarrow & & \downarrow \sigma_*\pi^* \\ D_c^+(X_{\bullet\mathrm{ét}}, A) & \longrightarrow & D_c^+(Y_{\bullet\mathrm{ét}}, A). \end{array}$$

Thus every finite type algebraic S-stack $\mathfrak{X}$ gives rise to a tractable cd-structure $D_c^+(\mathfrak{X}_{\mathrm{sm}}, A)$, and every morphism of finite type algebraic S-stacks defines a tractable morphism of these cd-structures. So we get a theory analogous to the ℓ-adic theory of Remark 4.75.

8. Purity and Extraordinary Pullbacks

DEFINITION 4.77. We call an ℓ-adic lcc sheaf M on the topos X *torsion free*, if for every $n \in \mathbb{N}$ the sheaf of Λ_n-modules M_n on X is locally free.

Let M be a torsion free ℓ-adic lcc sheaf on X. Then the A-linear functor
$$\_ \otimes_\Lambda M : \mathrm{Mod}(X^\mathbb{N}, \Lambda) \longrightarrow \mathrm{Mod}(X^\mathbb{N}, \Lambda)$$
is exact and maps AR-null modules to AR-null modules. If $(\mathcal{S}, \mathcal{L})$ is a closed L-stratification of X such that M is $(\mathcal{S}, \mathcal{L})$-constructible, then $\_ \otimes_\Lambda M$ also preserves $(\mathcal{S}, \mathcal{L})$-constructability. So if $(X, \overline{X})$ is a c-topos and M a torsion free ℓ-adic lcc sheaf on $\overline{X}$, then we get an induced functor
$$\_ \otimes_\Lambda M : \mathbb{D}_c(X, A) \longrightarrow \mathbb{D}_c(X, A),$$
at least if we assume that every L-stratification can be refined to a closed L-stratification.

Now let S be a scheme and n a natural number, invertible on S. Then μ_n, the S-scheme of n-th roots of unity, is finite étale over S, hence is an lcc sheaf of abelian groups on $S_{\mathrm{ét}}$. Let A be a discrete valuation ring with finite residue field of characteristic invertible over S and with parameter ℓ. Let l be the rational prime number that ℓ divides. Let us set
$$\mu_{\ell^{n+1}} = A \otimes_\mathbb{Z} \mu_{l^{n+1}}.$$

Then $(\mu_{\ell^{n+1}})_{n \in \mathbb{N}}$ is a torsion free ℓ-adic lcc sheaf on $S_{\mathrm{ét}}$ which we denote by $\Lambda(1)$. we will also denote by $\Lambda(1)$ the pullback to any algebraic S-stack. Similarly, we define
$$\mu_{\ell^{n+1}}^{\otimes m} = A \times_\mathbb{Z} \mu_{l^{n+1}}^{\otimes m},$$
for any $m \in \mathbb{Z}$ and denote the ℓ-adic sheaf $(\mu_{\ell^{n+1}}^{\otimes m})$ by $\Lambda(m)$.

DEFINITION 4.78. Let $\mathfrak{X}$ be an algebraic S-stack. For an object $M \in \mathrm{ob}\,\mathbb{D}_c(\mathfrak{X}_{\mathrm{sm}}, A)$ we write $M(1)$ for $M \otimes_\Lambda \Lambda(1)$, and call it the *Tate twist of* M. We define $M(n) = M \otimes_\Lambda \Lambda(n)$ and note that $M(n)(m) = M(n+m)$.

PROPOSITION 4.79 (Purity). *Let $\mathfrak{Y}$ be an algebraic S-stack of finite type. Let $\mathfrak{Z}$ and $\mathfrak{X}$ be algebraic S-stacks endowed with smooth finite type morphisms $f : \mathfrak{Z} \to \mathfrak{Y}$ and $g\mathfrak{X} \to \mathfrak{Y}$. Let $i : \mathfrak{Z} \to \mathfrak{X}$ be a closed immersion over $\mathfrak{Y}$. Let c be the codimension of $\mathfrak{Z}$ in $\mathfrak{X}$, which is a locally constant function on $|\mathfrak{Z}|$ with values in $\mathbb{N}$. Then for any object M of $\mathbb{D}^+_{\mathrm{lcc}}(\mathfrak{X}_{sm}, A)$ we have a canonical isomorphism*

$$i^*M(-c)[-2c] \xrightarrow{\sim} \mathbb{R}i^! M$$

in $\mathbb{D}^+_{\mathrm{lcc}}(\mathfrak{Z}_{sm}, A)$.

PROOF. Choose a presentation $Y \to \mathfrak{Y}$ of $\mathfrak{Y}$. Then let $X \to \mathfrak{X} \times_{\mathfrak{Y}} Y$ be a presentation, which induces a presentation $X \to \mathfrak{X}$ of X. Let $Z \to \mathfrak{Z}$ be the induced presentation of $\mathfrak{Z}$. Let $Z_\bullet$ and $X_\bullet$ be the simplicial algebraic spaces given by these presentations of $\mathfrak{Z}$ an $\mathfrak{X}$, respectively. Note that for every $p \in \mathbb{N}$ the scheme Z_p is a closed subscheme of X_p and $i_p : Z_p \to X_p$ is a smooth Y-pair of codimension c. Hence we have for any lcc sheaf F of A'-modules on $\mathfrak{X}_{\text{ét}}$ that

$$R^q i_p^!(F|X_p) = \begin{cases} 0 & \text{for } q \neq 2c \\ i_p^* F(-c) & \text{for } q = 2c. \end{cases} \quad (10)$$

We will work with fibered topoi over $\Delta_{\mathrm{op}} \times \mathbb{N}$.

We have a closed immersion of topoi

$$i : \mathrm{top}(Z_{\bullet\,\text{ét}})^{\mathbb{N}} \longrightarrow \mathrm{top}(X_{\bullet\,\text{ét}})^{\mathbb{N}}$$

as in Proposition 3.8. Hence we have a natural homomorphism

$$i^* R\mathcal{H}om(\Lambda, M) \longrightarrow R\mathcal{H}om(Ri^!\Lambda, Ri^! M) \quad (11)$$

for every $M \in \mathrm{ob}\, D^+(X_{\bullet\,\text{ét}}^{\mathbb{N}}, \Lambda)$. (This homomorphism is the adjoint of

$$R\mathcal{H}om(\Lambda, M) \longrightarrow i_* R\mathcal{H}om(Ri^!\Lambda, Ri^! M),$$

which is the derivative of

$$\mathcal{H}om(\Lambda, M) \longrightarrow i_* \mathcal{H}om(i^!\Lambda, i^! M).)$$

Now by (10), we have $Ri^!\Lambda = \Lambda(-c)[-2c]$, so the homomorphism (11) gives rise to

$$i^* M \longrightarrow Ri^! M(c)[2c].$$

Our goal is to prove that this is an isomorphism if $h^p M$ is lcc, for all p. But it suffices to do this after application of h^p, so it follows from (10), and the fact that $Ri^!$ and Tate twists may be calculated componentwise over $\Delta_{\mathrm{op}} \times \mathbb{N}$ (see Proposition 3.8). Note that, in particular, $Ri^! M$ is lcc. we get an induced isomorphism

$$i^* M \longrightarrow Ri^! M(c)[2c]$$

in $\mathbb{D}^+_{\mathrm{lcc}}(Z_{\bullet\,\text{ét}}, A) = \mathbb{D}^+_{\mathrm{lcc}}(\mathfrak{Z}_{\mathrm{sm}}, A)$. □

REMARK 4.80. Consider a 2-cartesian diagram

$$
\begin{array}{ccc}
\mathfrak{Z}' & \xrightarrow{i'} & \mathfrak{X}' \\
v \downarrow & \boxed{2} & \downarrow u \\
\mathfrak{Z} & \xrightarrow{i} & \mathfrak{X}
\end{array}
\qquad (12)
$$

of finite type algebraic S-stacks, where i is a closed immersion and u is smooth representable. Then we have

$$v^* \circ \mathbb{R}i^! = \mathbb{R}i'^! \circ u^*$$

as functors from $\mathbb{D}_c^+(\mathfrak{X}_{\mathrm{sm}}, A)$ to $\mathbb{D}_c^+(\mathfrak{Z}'_{\mathrm{sm}}, A)$. This is a formal consequence of the smooth base change theorem for $\mathbb{R}j_*$, where $j : \mathfrak{U} \to \mathfrak{X}$ is the open complement of $\mathfrak{Z}$ (see Corollary 4.74).

COROLLARY 4.81. *In the situation of Proposition 4.79 we have for every* $M \in \mathrm{ob}\,\mathbb{D}_c^+(\mathfrak{Y}_{sm}, A)$ *that*

$$f^* M(-c)[-2c] = \mathbb{R}i^! g^* M.$$

PROOF. Choose an L-stratification $(\mathcal{S}, \mathcal{L})$ of $\mathfrak{Y}$ such that $M \in \mathrm{ob}\,\mathbb{D}_{c,(\mathcal{S},\mathcal{L})}^+(\mathfrak{Y}_{\mathrm{sm}}, A)$. As in the proof of the proposition we get a canonical homomorphism

$$f^* M(-c)[-2c] \longrightarrow \mathbb{R}i^! g^* M.$$

To prove it is an isomorphism, we may pass to the strata of $\mathfrak{Y}$ using extraordinary pullbacks, by Remark 4.80. Then we are reduced to the case considered in the proposition. □

DEFINITION 4.82. We call a morphism $f : \mathfrak{X} \to \mathfrak{Y}$ of finite type algebraic S-stacks an *elementary embeddable morphism*, if f is a closed immersion, a representable universal homeomorphism or a smooth representable morphism. We call f *embeddable*, if it may be factored into a finite number of elementary embeddable morphisms.

Note that, by definition, embeddable morphisms are representable. Compositions and base changes of embeddable morphisms are embeddable.

PROPOSITION 4.83. *Let $\mathfrak{X}$ be an algebraic S-stack of finite type and $x : \mathrm{Spec}\, k \to \mathfrak{X}$ a finite type point of $\mathfrak{X}$. Then x is embeddable.*

PROOF. Using similar tricks as in the proof of the second devissage lemma (Proposition 4.17) we reduce to the following considerations. If $X \to \mathrm{Spec}\, k$ is a scheme, then any section is a locally closed immersion, hence embeddable. Let G be a smooth group scheme over k. Then $\mathrm{Spec}\, k \to BG$ is smooth and representable, hence embeddable. □

DEFINITION 4.84. Let $f : \mathfrak{X} \to \mathfrak{Y}$ be an embeddable morphism of algebraic S-stacks. Then we define

$$\mathbb{R}f^! : \mathbb{D}_c^+(\mathfrak{Y}_{\mathrm{sm}}, A) \longrightarrow \mathbb{D}_c^+(\mathfrak{X}_{\mathrm{sm}}, A)$$

as follows. If f is a closed immersion, $\mathbb{R}f^!$ is any right adjoint of $\mathbb{R}f_*$, which exists by Definition 4.53. If f is smooth, then we set $\mathbb{R}f^! = f^*(d)[2d]$, where d is the relative dimension of f, which is a locally constant $\mathbb{N}$-valued function on $|\mathfrak{X}|$. If f is a universal homeomorphism we set $\mathbb{R}f^! = f^*$. Finally, in the general case, we factor f into elementary embeddable morphisms and compose the above definitions for these.

PROPOSITION 4.85. *This definition makes sense.*

PROOF. We need to show that if we factor f in two ways into elementary embeddable morphisms, we arrive at the same $\mathbb{R}f^!$-functor. We will show that this results from the following two facts.
  (1) In the situation of Proposition 4.79, where f and g are smooth representable and i is a closed immersion, we have $\mathbb{R}i^!\mathbb{R}g^! = \mathbb{R}f^!$.
  (2) Consider a 2-cartesian diagram (12) as in Remark 4.80. Then $\mathbb{R}v^!\mathbb{R}i^! = \mathbb{R}i'^!\mathbb{R}u^!$.

Use (1) and (2) to prove that Claim (2) remains valid if we replace the 2-cartesian requirement by mere commutativity of the diagram. Deduce also that we have the truth of (1), in the case where f is a closed immersion instead of a smooth map. From this fact we get also that if $g \circ j = f \circ i$, where i and j are closed immersions and f and g are smooth representable, then $\mathbb{R}i^!\mathbb{R}f^! = \mathbb{R}j^!\mathbb{R}g^!$.

Now it is easy to prove the following fact. Consider an equation $f \circ i = j$, where i and j are locally closed immersions an f elementary embeddable. Then there exist factorizations of i and j into closed immersions followed by open immersions such that for the induced definitions of $\mathbb{R}i^!$ and $\mathbb{R}j^!$ we have $\mathbb{R}i^!\mathbb{R}f^! = \mathbb{R}j^!$. From this we deduce, using the fact that any section of a representable morphism is a locally closed immersion, that no matter how we factor $\mathrm{id}_\mathfrak{X}$ into elementary embeddable morphisms, $\mathbb{R}\,\mathrm{id}^!_\mathfrak{X} = \mathrm{id}$. The general case of an embeddable morphism $f: \mathfrak{X} \to \mathfrak{Y}$ now follows by applying this to the factorization $\mathrm{id}_\mathfrak{X} = \mathrm{pr}_1 \circ \Delta_f$, and using (2).

So it remains to prove (1) and (2). Note that (2) is a trivial consequence of Remark 4.80 and (1) follows easily from Corollary 4.81. □

PROPOSITION 4.86 (Base Change). *Let $f: \mathfrak{X} \to \mathfrak{Y}$ be a morphism of finite type algebraic S-stacks. Let $j: \mathfrak{Y}' \to \mathfrak{Y}$ be embeddable. Then there is an isomorphism of functors $\mathbb{R}u^!\mathbb{R}f_* \to \mathbb{R}g_*\mathbb{R}v^!$ from $\mathbb{D}^+_c(\mathfrak{X}_{sm}, A)$ to $\mathbb{D}^+_c(\mathfrak{Y}'_{sm}, A)$, where the notations are as in the diagram*

$$\begin{array}{ccc} \mathfrak{X}' & \xrightarrow{g} & \mathfrak{Y}' \\ v \downarrow & \boxed{2} & \downarrow u \\ \mathfrak{X} & \xrightarrow{f} & \mathfrak{Y}. \end{array}$$

PROOF. Without loss of generality u is elementary embeddable. If u is smooth this is the smooth base change theorem Corollary 4.74. If u is a closed immersion, it is equivalent to the trivial base change to the open complement of $\mathfrak{Y}'$ in $\mathfrak{Y}$. □

PROPOSITION 4.87. *Assuming that S is the spectrum of a field, let $\mathfrak{X}$ be smooth over S of dimension n and let $M \in \mathrm{ob}\, \mathbb{D}^+_{\mathrm{lcc}}(\mathfrak{X}, A)$. If $x : \mathrm{Spec}\, k \to \mathfrak{X}$ is a finite type point of $\mathfrak{X}$, then we have $\mathbb{R}x^! M = x^* M(-n)[-2n]$.*

PROOF. This works similarly as Proposition 4.83. □

CHAPTER 5

Convergent Complexes

1. $\overline{Q}_\ell$-Complexes

Let S be a scheme of finite type over a noetherian regular scheme of dimension zero or one and let ℓ be a prime number invertible on S. In the previous section, we constructed for every algebraic S-stack $\mathfrak{X}$ of finite type and every discrete valuation ring A with residue characteristic ℓ an A-t-category $\mathbb{D}_c(\mathfrak{X}_{\mathrm{sm}}, A)$ whose heart is $\mathrm{Mod}_c(\mathfrak{X}_{\mathrm{\acute{e}t}}, A)$. Let K be the quotient field of A. By extension of scalars we get a K-t-category $\mathbb{D}_c(\mathfrak{X}_{\mathrm{sm}}, K)$ whose heart is $\mathrm{Mod}_c(\mathfrak{X}_{\mathrm{\acute{e}t}}, K)$.

Taking the 2-limit over all finite extensions K of $\mathbb{Q}_\ell$, we get a $\overline{Q}_\ell$-t-category $\mathbb{D}_c(\mathfrak{X}_{\mathrm{sm}}, \overline{Q}_\ell)$ whose heart is $\mathrm{Mod}_c(\mathfrak{X}_{\mathrm{\acute{e}t}}, \overline{Q}_\ell)$. We call the objects of $\mathbb{D}_c(\mathfrak{X}_{\mathrm{sm}}, \overline{Q}_\ell)$ *constructible $\overline{Q}_\ell$-complexes on* $\mathfrak{X}$ and the objects of $\mathrm{Mod}_c(\mathfrak{X}_{\mathrm{\acute{e}t}}, \overline{Q}_\ell)$ *constructible $\overline{Q}_\ell$-sheaves on* $\mathfrak{X}$.

By abuse of notation, we denote by A also the object of $\mathrm{Mod}_c(\mathfrak{X}_{\mathrm{\acute{e}t}}, A)$ defined by the sheaf of rigs $\Lambda = (\Lambda_n)_{n \in \mathbb{N}}$ on $X^{\mathbb{N}}$, where $\Lambda_n = A/\mathfrak{l}^{n+1}$ and $\mathfrak{l}$ is the maximal ideal of A. The image in $\mathrm{Mod}_c(\mathfrak{X}_{\mathrm{\acute{e}t}}, K)$ will be denoted by K, the image in $\mathrm{Mod}_c(\mathfrak{X}_{\mathrm{\acute{e}t}}, \overline{Q}_\ell)$ by $\overline{Q}_\ell$ or simply $\overline{\mathbb{Q}}_\ell$.

DEFINITION 5.1. We call a constructible $\overline{Q}_\ell$-sheaf F on $\mathfrak{X}$ *lisse*, if it may be represented by an object of $\mathrm{Mod}_{\mathrm{lcc}}(\mathfrak{X}_{\mathrm{\acute{e}t}}, A)$, for the ring of integers A in some finite extension of $\mathbb{Q}_\ell$.

PROPOSITION 5.2. *Every lisse $\overline{Q}_\ell$-sheaf is torsion-free (see Definition 4.77).*

PROOF. Let A be the ring of integers in a finite extension of $\mathbb{Q}_\ell$ and let $F = (F_n)_{n \in \mathbb{N}}$ be an object of $\mathrm{Mod}_{\mathrm{lcc}}(\mathfrak{X}_{\mathrm{\acute{e}t}}, A)$. Without loss of generality we may assume that for every $r \geq 1$ the kernel of multiplication by ℓ^r on F is AR-null. (We have tacitly changed notation, to denote by ℓ a parameter of A.) We wish to see that for every pair of non-negative integers (n, r) such that $n + 1 \geq r$ we have an exact sequence

$$0 \longrightarrow \ell^{n-r+1} F_n \longrightarrow F_n \xrightarrow{\ell^r} F_n.$$

This reduces to a (not very difficult) question of ℓ-adic algebra, since $\mathfrak{X}_{\mathrm{\acute{e}t}}$ has sufficiently many points. □

COROLLARY 5.3 (Projection Formula). *Let $f : \mathfrak{X} \to \mathfrak{Y}$ be a smooth morphism of finite type algebraic S-stacks and F a lisse $\overline{Q}_\ell$-sheaf on $\mathfrak{Y}$.*

Then for every $M \in \operatorname{ob}\operatorname{Mod}_c(\mathfrak{X}_{\text{ét}}, \mathbb{Q}_\ell)$ we have

$$R^n f_*(f^*F \otimes M) = F \otimes R^n f_* M,$$

for all $n \geq 0$.

PROOF. This follows from the general projection formula for morphisms of ringed topoi noting that a torsion free object of $\operatorname{Mod}_{\text{lcc}}(\mathfrak{Y}_{\text{ét}}, A)$ gives rise to a locally free sheaf of Λ-modules on $\mathfrak{Y}_{\text{ét}}^{\mathbb{N}}$. $\square$

For the remainder of this section, let us assume for simplicity that S is the spectrum of a field k. For a finite type algebraic k-stack $f : \mathfrak{X} \to S$ and a constructible $\mathbb{Q}_\ell$-complex M on $\mathfrak{X}$ we abbreviate $R^n f_* M$ by $H^n(\mathfrak{X}, M)$, for all $n \geq 0$. We may consider $H^n(\mathfrak{X}, M)$ as a $\mathbb{Q}_\ell$-vector space with $\operatorname{Gal}(k)$-action.

Let G be a smooth connected group scheme over S. Let X be an S-scheme of finite type on which G acts. Let $f : \mathfrak{X} \to \mathfrak{Y}$ be a fiber bundle with structure group G and fiber X. This means that there exists a $\mathfrak{Y}$-space $\mathfrak{P}$, which is a principal homogeneous G-bundle over $\mathfrak{Y}$, such that $\mathfrak{P} \times_G X \cong \mathfrak{X}$ as $\mathfrak{Y}$-spaces. Use notation as in the diagram

$$\begin{array}{ccc} \mathfrak{X} & \xrightarrow{f} & \mathfrak{Y} \\ {\scriptstyle \tau} \searrow & & \downarrow {\scriptstyle \rho} \\ & S. & \end{array}$$

LEMMA 5.4. *We have* $R^n f_* \mathbb{Q}_\ell \cong \rho^* H^n(X, \mathbb{Q}_\ell)$, *for all* $n \geq 0$.

PROOF. We need to show that $R^n f_* \mathbb{Z}/\ell^{\nu+1} = \rho^* H^n(X, \mathbb{Z}/\ell^{\nu+1})$, for all ν. By Lemma 1.4.1 of [**3**] we have

$$\begin{aligned} R^n f_* \mathbb{Z}/\ell^{\nu+1} &= \mathfrak{P} \times_G \rho^* H^n(X, \mathbb{Z}/\ell^{\nu+1}) \\ &= \rho^* H^n(X, \mathbb{Z}/\ell^{\nu+1}) \end{aligned}$$

since G is connected. $\square$

PROPOSITION 5.5. *We have for every lisse $\mathbb{Q}_\ell$-sheaf F on $\mathfrak{Y}$ a spectral sequence*

$$H^p(\mathfrak{Y}, F) \otimes H^q(X, \mathbb{Q}_\ell) \Longrightarrow H^{p+q}(\mathfrak{X}, f^*F).$$

PROOF. This is the Leray spectral sequence of the composition $\tau = \rho \circ f$. Let us check the form of the E_2-term. Indeed, we have

$$\begin{aligned} H^p(\mathfrak{Y}, R^q f_* f^* F) &= H^p(\mathfrak{Y}, F \otimes R^q f_* \mathbb{Q}_\ell) \\ &= H^p(\mathfrak{Y}, F \otimes \rho^* H^q(X, \mathbb{Q}_\ell)) \\ &= H^p(\mathfrak{Y}, F) \otimes H^q(X, \mathbb{Q}_\ell)), \end{aligned}$$

by the lemma and the projection formula. $\square$

1.1. A Theorem of Borel.
We need to translate the main theorem from [6] into our context.

THEOREM 5.6. *Let k be a field and G/k a connected group variety. consider the spectral $\mathbb{Q}_\ell$-algebra given by the universal fibration $\operatorname{Spec} k \to BG$,*

$$E_2^{p,q} = H^p(BG, \mathbb{Q}_\ell) \otimes_{\mathbb{Q}_\ell} H^q(G, \mathbb{Q}_\ell) \Longrightarrow \mathbb{Q}_\ell. \tag{13}$$

For every $q \geq 0$ the $\mathbb{Q}_\ell$-vector space $E_r^{0,q}$ is a subspace of $H^q(G, \mathbb{Q}_\ell)$, for all $r \geq 2$. In particular, we have the transgressive subspace

$$N^p = E_{q+1}^{0,q} \subset H^p(G, \mathbb{Q}_\ell),$$

for $q \geq 1$. Consider the graded $\mathbb{Q}_\ell$-vector space $N = \bigoplus_{q \geq 1} N^q$. We have
(1) *$N^q = 0$ if q is even.*
(2) *The canonical map $\Lambda N \to H^*(G, \mathbb{Q}_\ell)$ is an isomorphism of graded $\mathbb{Q}_\ell$-algebras.*
(3) *The spectral sequence (13) induces an epimorphism of graded $\mathbb{Q}_\ell$-vector spaces $H^*(BG, \mathbb{Q}_\ell) \to N[-1]$. Any section induces an isomorphism $S(N[-1]) \xrightarrow{\sim} H^*(BG, \mathbb{Q}_\ell)$.*

PROOF. This is Théorème 13.1 from [6]. □

REMARK 5.7. If the transgression N is homogeneous, then $H^*(BG, \mathbb{Q}_\ell) = S(N[-1])$, canonically. So in this case the theorem holds over an arbitrary base S. For example, let $f : \mathfrak{E} \to \mathfrak{X}$ be a family of elliptic curves over the finite type algebraic S-stack $\mathfrak{X}$ and consider the morphism $\pi : B(\mathfrak{E}/\mathfrak{X}) \to \mathfrak{X}$. If we denote $\mathbb{R}^1 f_* \mathbb{Q}_\ell$ by $H^1(\mathfrak{E})$, then we have $\mathbb{R}^{2n} \pi_* \mathbb{Q}_\ell = S^n H^1(\mathfrak{E})$ and $\mathbb{R}^{2n+1} \pi_* \mathbb{Q}_\ell = 0$, for all $n \in \mathbb{N}$.

2. Frobenius

Let circ denote the topos we already encountered in Example 3.31. The t-category $\mathbb{D}_c(\text{circ}, Q_\ell)$ has heart $\operatorname{Mod}_c(\text{circ}, Q_\ell)$, which is the category of automorphisms of finite dimensional Q_ℓ-vector spaces whose eigenvalues are ℓ-adic units. We will denote objects of $\operatorname{Mod}_c(\text{circ}, Q_\ell)$ by (M, f).

Let $q \in \mathbb{R}_{>0}$. Recall from Définition 1.2.1 in [9] that a number is called *pure of weight $n \in \mathbb{Z}$ relative q*, if its absolute value is $\sqrt{q}^n$, no matter how it is embedded in $\mathbb{C}$.

DEFINITION 5.8. We call an object $(M, f) \in \operatorname{ob} \operatorname{Mod}_c(\text{circ}, Q_\ell)$ *pure of weight $n \in \mathbb{Z}$ relative q*, if every eigenvalue of f on M is pure of weight n relative q. We call (M, f) *mixed relative q*, if every eigenvalue of f on M is pure of some integer weight relative q. Finally, an object $M \in \operatorname{ob} \mathbb{D}_c(\text{circ}, Q_\ell)$ is called *mixed relative q*, if for every $i \in \mathbb{Z}$ the object $h^i(M)$ is mixed relative q.

NOTE 5.9. Let us denote by $\operatorname{Mod}_{m_q}(\text{circ}, Q_\ell)$ the full subcategory of $\operatorname{Mod}_c(\text{circ} \, Q_\ell)$ of mixed objects relative q. Then this is a closed subcategory. Thus $\mathbb{D}_{m_q}(\text{circ}, Q_\ell)$, the full subcategory of $\mathbb{D}_c(\text{circ}, Q_\ell)$ consisting of mixed objects relative q, is a t-category with heart $\text{Mod}_{m_q}(\text{circ}, Q_\ell)$. Note that for any $n \in \mathbb{Z}$ we have $\text{Mod}_{m_{q^n}}(\text{circ}, Q_\ell) \subset \text{Mod}_{m_q}(\text{circ}, Q_\ell)$ and thus $\mathbb{D}_{m_{q^n}}(\text{circ}, Q_\ell) \subset \mathbb{D}_{m_q}(\text{circ}, Q_\ell)$.

Let $M \in \text{ob}\,\text{Mod}_{m_q}(\text{circ}, Q_\ell)$ be a mixed automorphism of a finite dimensional Q_ℓ-vector space. Let

$$M = \bigoplus_{p \in \mathbb{Z}} \text{gr}_p^{w(q)} M$$

be the decomposition into pure factors, $\text{gr}_p^{w(q)} M$ being pure of weight p relative q.

DEFINITION 5.10. We call the object $M \in \text{ob}\,\mathbb{D}_{m_q}^+(\text{circ}, Q_\ell)$ *absolutely convergent* (or simply *convergent*), if

$$w_q(M)(z) = \sum_p \sum_i \dim(\text{gr}_p^{w(q)} h^i M) z^p$$

converges uniformly on compact subsets of the punctured unit disc $0 < |z| < 1$, and has at worst a pole at the origin $z = 0$. Let us denote the full subcategory of $\mathbb{D}_{m_q}^+(\text{circ}, Q_\ell)$ consisting of absolutely convergent objects by $\mathbb{D}_{a_q}^+(\text{circ}, Q_\ell)$.

PROPOSITION 5.11. *The category $\mathbb{D}_{a_q}^+(\text{circ}, Q_\ell)$ is a triangulated subcategory of $\mathbb{D}_{m_q}^+(\text{circ}, Q_\ell)$. It is a Q_ℓ-t-category with heart $\text{Mod}_{m_q}(\text{circ}, Q_\ell)$. Moreover, we have*

$$\mathbb{D}_{m_{q^n}}^+(\text{circ}, Q_\ell) \cap \mathbb{D}_{a_q}^+(\text{circ}, Q_\ell) = \mathbb{D}_{a_{q^n}}^+(\text{circ}, Q_\ell),$$

as subcategories of $\mathbb{D}_{m_q}^+(\text{circ}, Q_\ell)$, if $n > 0$.

Consider the automorphism $\epsilon_n : \text{circ} \to \text{circ}$. Recall that $\epsilon_n^*(M, f) = (M, f^n)$, for an object (M, f) of circ or $\text{Mod}_c(\text{circ}, Q_\ell)$. So ϵ_n^* multiplies weights relative q by n. Thus we have that $M \in \text{ob}\,\text{Mod}_c(\text{circ}, Q_\ell)$ is mixed relative q if and only if $\epsilon_n^* M$ is mixed relative q^n. The same is true for objects of $\mathbb{D}_c^+(\text{circ}, Q_\ell)$. In particular, we have

$$\epsilon_n^* : \mathbb{D}_{m_q}^+(\text{circ}, Q_\ell) \longrightarrow \mathbb{D}_{m_{q^n}}^+(\text{circ}, Q_\ell) \subset \mathbb{D}_{m_q}^+(\text{circ}, Q_\ell).$$

Now assume that $n > 0$. Let $M \in \text{ob}\,\mathbb{D}_{m_q}^+(\text{circ}, Q_\ell)$. We have $w_{q^n}(\epsilon_n^* M)(z) = w_q(M)(z)$. So M is convergent if and only if $\epsilon_n^* M$ is. In particular, we get

$$\epsilon_n^* : \mathbb{D}_{a_q}^+(\text{circ}, Q_\ell) \longrightarrow \mathbb{D}_{a_{q^n}}^+(\text{circ}, Q_\ell) \subset \mathbb{D}_{a_q}^+(\text{circ}, Q_\ell).$$

Let N be an object of $\text{Mod}_c(\text{circ}, Q_\ell)$. Assume that N is mixed relative q. Then the functor

$$\_ \otimes_{Q_\ell} N : \mathbb{D}_c^+(\text{circ}, Q_\ell) \longrightarrow \mathbb{D}_c^+(\text{circ}, Q_\ell)$$

preserves mixedness relative q. Let $M \in \mathrm{ob}\,\mathbb{D}^+_{m_q}(\mathrm{circ}, Q_\ell)$. We have $w_q(M \otimes N) = w_q(M) w_q(N)$, and since $w_q(N)$ is polynomial, $\_ \otimes_{Q_\ell} N$ preserves convergence.

Now let $\mathbb{F}_q$ be a finite field such that $\ell \nmid q$. The choice of an algebraic closure $\overline{\mathbb{F}}_q$ of $\mathbb{F}_q$ determines an equivalence

$$\begin{aligned}(\mathrm{Spec}\,\mathbb{F}_q)_{\text{ét}} &\longrightarrow \mathrm{circ} \\ X &\longmapsto (X(\overline{\mathbb{F}}_q), \phi),\end{aligned} \qquad (14)$$

where $\phi : X(\overline{\mathbb{F}}_q) \to X(\overline{\mathbb{F}}_q)$ is the map given by $\phi(x) = x \circ \mathrm{Spec}(\phi_q)$, where $\phi_q : \overline{\mathbb{F}}_q \to \overline{\mathbb{F}}_q$ is the Frobenius $\phi_q(\alpha) = \alpha^q$. A different choice of algebraic closure of $\mathbb{F}_q$ gives rise to an isomorphic equivalence of topoi.

The equivalence (14) induces an equivalence $\mathbb{D}^+_c(\mathrm{Spec}\,\mathbb{F}_q, Q_\ell) \to \mathbb{D}^+_c(\mathrm{circ}, Q_\ell)$. Again, a different choice of algebraic closure gives rise to an isomorphic equivalence.

DEFINITION 5.12. We call an object M of $\mathbb{D}^+_c(\mathrm{Spec}\,\mathbb{F}_q, Q_\ell)$ *mixed (convergent) relative* $q_0 \in \mathbb{R}_{>0}$, if any corresponding object of $\mathbb{D}^+_c(\mathrm{circ}, Q_\ell)$ is mixed (convergent) relative $1/q_0$. We say M is *mixed (convergent)*, if M is mixed (convergent) relative q.

Thus we get categories $\mathbb{D}^+_{m_{q_0}}(\mathrm{Spec}\,\mathbb{F}_q, Q_\ell)$ and $\mathbb{D}^+_{a_{q_0}}(\mathrm{Spec}\,\mathbb{F}_q, Q_\ell)$, for any $q_0 \in \mathbb{R}_{>0}$. We set $\mathbb{D}^+_m(\mathrm{Spec}\,\mathbb{F}_q, Q_\ell) = \mathbb{D}^+_{m_q}(\mathrm{Spec}\,\mathbb{F}_q, Q_\ell)$ and $\mathbb{D}^+_a(\mathrm{Spec}\,\mathbb{F}_q, Q_\ell) = \mathbb{D}^+_{a_q}(\mathrm{Spec}\,\mathbb{F}_q, Q_\ell)$.

Finally, we call an ℓ-adic sheaf $F \in \mathrm{ob}\,\mathrm{Mod}_m(\mathrm{Spec}\,\mathbb{F}_q, Q_\ell)$ *pure of weight* n, if any corresponding ℓ-adic sheaf on circ is pure of weight n relative $1/q$.

If $M \in \mathrm{ob}\,\mathbb{D}^+_a(\mathrm{Spec}\,\mathbb{F}_q, Q_\ell)$, then the function

$$w(M)(z) = \sum_p \sum_i \dim(\mathrm{gr}^w_p h^i M) z^p$$

is meromorphic on the unit disc in $\mathbb{C}$.

Let $\mathbb{F}_q \to \mathbb{F}_{q^n}$ be a homomorphism of finite fields. Let $f : \mathrm{Spec}\,\mathbb{F}_{q^n} \to \mathrm{Spec}\,\mathbb{F}_q$ denote the corresponding morphism of schemes. We get an induced morphism $(\mathrm{Spec}\,\mathbb{F}_{q^n})_{\text{ét}} \to (\mathrm{Spec}\,\mathbb{F}_q)_{\text{ét}}$ of topoi, which induces via (14) (choosing a common algebraic closure of $\mathbb{F}_q$ and $\mathbb{F}_{q^n}$) the morphism $\epsilon_n : \mathrm{circ} \to \mathrm{circ}$. So the functor $f^* : \mathbb{D}^+_c(\mathrm{Spec}\,\mathbb{F}_q, Q_\ell) \to \mathbb{D}^+_c(\mathrm{Spec}\,\mathbb{F}_{q^n}, Q_\ell)$ has the property that $M \in \mathrm{ob}\,\mathbb{D}^+_c(\mathrm{Spec}\,\mathbb{F}_q, Q_\ell)$ is mixed relative q_0 if and only if f^*M is mixed relative q_0^n. In particular, M is mixed if and only if f^*M is. Moreover, $M \in \mathrm{ob}\,\mathbb{D}^+_m(\mathrm{Spec}\,\mathbb{F}_q, Q_\ell)$ is convergent if and only if f^*M is convergent. Finally, $F \in \mathrm{ob}\,\mathrm{Mod}_c(\mathrm{Spec}\,\mathbb{F}_q, Q_\ell)$ is pure of weight n if and only if f^*F is.

We also have $M \in \mathrm{ob}\,\mathbb{D}^+_c(\mathrm{Spec}\,\mathbb{F}_q, Q_\ell)$ is mixed if and only if $M(n)$ is mixed and $M \in \mathrm{ob}\,\mathbb{D}^+_m(\mathrm{Spec}\,\mathbb{F}_q, Q_\ell)$ is convergent if and only if $M(n)$ is convergent.

3. Mixed and Convergent Complexes

Let ℓ be a prime number, S a scheme of finite type over $\mathbb{Z}[1/\ell]$ and $\mathfrak{X}$ a finite type algebraic S-stack.

DEFINITION 5.13. We call an ℓ-adic sheaf $F \in \operatorname{ob} \operatorname{Mod}_c(\mathfrak{X}, Q_\ell)$ *punctually pure of weight* $n \in \mathbb{Z}$, if for every finite field $\mathbb{F}_q$ and every S-morphism $x : \operatorname{Spec} \mathbb{F}_q \to \mathfrak{X}$, the pullback x^*F is pure of weight n.

NOTE 5.14. For $\mathfrak{X} = \mathbb{F}_q$ the notion of punctual purity coincides with purity. This definition is equivalent to demanding that for every finite type point x of $\mathfrak{X}$, choosing a representative $x : \operatorname{Spec} \mathbb{F}_q \to \mathfrak{X}$ of x, the pullback x^*F is pure of weight n. If $f : \mathfrak{X} \to \mathfrak{Y}$ is a finite type morphism of algebraic S-stacks and F is a punctually pure ℓ-adic sheaf on $\mathfrak{Y}$, then f^*F is punctually pure of the same weight.

If $\mathfrak{X}$ is a scheme, this definition coincides with Definition 1.2.2 in [**9**]. We work with the arithmetic Frobenius whereas Deligne uses the geometric Frobenius in [loc. cit.]. This discrepancy is made up for by the switch from q_0 to $1/q_0$ in Definition 5.12.

PROPOSITION 5.15. *Let $f : X \to \mathfrak{X}$ be a finite type presentation of $\mathfrak{X}$. Then an ℓ-adic sheaf $F \in \operatorname{ob} \operatorname{Mod}_c(\mathfrak{X}, Q_\ell)$ is punctually pure of weight n if and only if f^*F is.*

PROOF. This follows from the fact that if $x : \operatorname{Spec} \mathbb{F}_q \to \mathfrak{X}$ is a given morphism, then there exists a finite extension $\mathbb{F}_{q^n}$ of $\mathbb{F}_q$ such that the induced morphism $\operatorname{Spec} \mathbb{F}_{q^n} \to \mathfrak{X}$ lifts to X. $\square$

DEFINITION 5.16. We call a constructible Q_ℓ-complex $M \in \operatorname{ob} \mathbb{D}_c^+(\mathfrak{X}, Q_\ell)$ on $\mathfrak{X}$ *mixed*, if for every $i \in \mathbb{Z}$, the ℓ-adic sheaf $h^i M$ has a filtration whose factors are punctually pure. Denote by $\mathbb{D}_m^+(\mathfrak{X}, Q_\ell)$ the full subcategory of $\mathbb{D}_c^+(\mathfrak{X}, Q_\ell)$ consisting of mixed objects and let $\operatorname{Mod}_m(\mathfrak{X}, Q_\ell) = \operatorname{Mod}_c(\mathfrak{X}, Q_\ell) \cap \mathbb{D}_m^+(\mathfrak{X}, Q_\ell)$.

PROPOSITION 5.17. *Let $f : X \to \mathfrak{X}$ be a finite type presentation of $\mathfrak{X}$. An object $M \in \operatorname{ob} \mathbb{D}_c^+(\mathfrak{X}, Q_\ell)$ is mixed if and only if f^*M is.*

PROOF. This follows from descent for ℓ-adic sheaves and Théorème 3.4.1(ii) in [**9**], which says that the weight filtration is canonical and so it descends. $\square$

PROPOSITION 5.18. *The category $\operatorname{Mod}_m(\mathfrak{X}, Q_\ell)$ is a closed subcategory of $\operatorname{Mod}_c(\mathfrak{X}, Q_\ell)$ and $\mathbb{D}_m^+(\mathfrak{X}, Q_\ell)$ is a sub-t-category of $\mathbb{D}_c^+(\mathfrak{X}, Q_\ell)$ whose heart is $\operatorname{Mod}_m(\mathfrak{X}, Q_\ell)$.*

PROOF. We need to show that the kernel and cokernel of a homomorphism of mixed sheaves are mixed. by Proposition 5.17 we may assume that $\mathfrak{X}$ is a scheme. Then this follows from the fact that every homomorphism respects the weight filtration. $\square$

PROPOSITION 5.19. *Assume that S lies over a field. Let $f : \mathfrak{X} \to \mathfrak{Y}$ be a finite type morphism of algebraic S-stacks. If $M \in \operatorname{ob} \mathbb{D}_c^+(\mathfrak{X}, Q_\ell)$ is mixed, then so is $\mathbb{R}f_*M$. If f is embeddable and $M \in \operatorname{ob} \mathbb{D}_c^+(\mathfrak{Y}, Q_\ell)$ is mixed, then so is $\mathbb{R}f^!M$.*

PROOF. Let us first assume that $f : X \to Y$ is a morphism of schemes. Let $M \in \operatorname{ob} \mathbb{D}_m^+(X, Q_\ell)$. To prove that $\mathbb{R}f_*M$ is mixed, we may assume that M is a mixed Q_ℓ-sheaf. Then M is a mixed Q_ℓ-sheaf in the sense of [9] and $\mathbb{R}^i f_*M$ is the $R^i f_*M$ considered in [loc. cit.]. Thus by Théorème 6.1.2 of [loc. cit.] $\mathbb{R}f_*M$ is indeed mixed.

By Proposition 5.17 and smooth base change we have thus proved that if f is an open immersion, then $\mathbb{R}f_*$ preserves mixedness. This immediately implies that $\mathbb{R}f^!$ preserves mixedness if f is a closed immersion. Consequently, $\mathbb{R}f^!$ preserves mixedness for any embeddable morphism f. This in turn implies that the property '$\mathbb{R}f_*$ preserves mixedness' satisfies Condition (2) of the second devissage lemma (Proposition 4.17). Thus by obvious amplifications of the second devissage lemma, we reduce to considering the case that $\mathfrak{Y} = Y$ is a scheme and $\mathfrak{X} = B(G/Y)$, where G is a smooth group scheme over Y. Then using the spectral sequence of the covering $Y \to \mathfrak{X}$, we reduce to the case of schemes. □

COROLLARY 5.20. *Assume that S lies over a field and that $\mathfrak{X}$ is a finite type algebraic S-stack. Then $\mathfrak{V} \mapsto \mathbb{D}_m^+(\mathfrak{V}, Q_\ell)$, where $\mathfrak{V}$ runs over the locally closed substacks of $\mathfrak{X}$, is as d-structure on $|\mathfrak{X}|$.*

Every morphism of finite type algebraic S-stacks induces a morphism of d-structures.

3.1. Convergence. Let S be a scheme of finite type over $\mathbb{F}_q$, where $\ell \nmid q$.

DEFINITION 5.21. A mixed Q_ℓ-complex $M \in \operatorname{ob} \mathbb{D}_m^+(\mathfrak{X}, Q_\ell)$ is called *absolutely convergent* (or simply *convergent*), if for every extension $\mathbb{F}_{q^n}$ of $\mathbb{F}_q$ and every S-morphism $x : \operatorname{Spec} \mathbb{F}_{q^n} \to \mathfrak{X}$, the extraordinary pullback $\mathbb{R}x^!M$ is absolutely convergent.

NOTE 5.22. This definition does not conflict with the earlier definition for $\mathfrak{X} = \operatorname{Spec} \mathbb{F}_q$. It is equivalent to demanding that for every finite type point x of $\mathfrak{X}$, choosing a representative $x : \operatorname{Spec} \mathbb{F}_q \to \mathfrak{X}$ of x, the extraordinary pullback $\mathbb{R}x^!M$ is convergent. If $f : \mathfrak{X} \to \mathfrak{Y}$ is an embeddable morphism of finite type algebraic S-stacks, then for any convergent object $M \in \operatorname{ob} \mathbb{D}_m^+(\mathfrak{X}, Q_\ell)$ the extraordinary pullback $\mathbb{R}f^!M$ is convergent. If $f : \mathfrak{X} \to \mathfrak{Y}$ is a smooth representable epimorphism and $M \in \operatorname{ob} \mathbb{D}_m^+(\mathfrak{Y}, Q_\ell)$, then M is convergent if and only if $\mathbb{R}f^!M$ is convergent and if and only if f^*M is convergent.

REMARK 5.23. There is of course another definition of convergence suggesting itself, namely the definition using ordinary pullbacks. It is not obvious that the two definitions should be equivalent. A related problem is the question to what extent convergence is stable under ordinary pullbacks.

DEFINITION 5.24. Let us denote the full subcategory of $\mathbb{D}_m^+(\mathfrak{X}, Q_\ell)$ consisting of absolutely convergent complexes by $\mathbb{D}_a^+(\mathfrak{X}, Q_\ell)$.

NOTE 5.25. The category $\mathbb{D}_a^+(\mathfrak{X}, Q_\ell)$ is a triangulated subcategory of $\mathbb{D}_m^+(\mathfrak{X}, Q_\ell)$. Every mixed Q_ℓ-sheaf is convergent (as is any bounded mixed complex). So $\mathbb{D}_a^+(\mathfrak{X}, Q_\ell)$ is a Q_ℓ-t-category with heart $\mathrm{Mod}_m(\mathfrak{X}, Q_\ell)$.

Before we can prove our main theorem, we need one more auxiliary result.

LEMMA 5.26. *Let X be an algebraic variety over the algebraically closed field k, and let A be a discrete valuation ring with finite residue field of characteristic invertible in k. Let $\mathcal{L} = (L_1, \ldots, L_n)$ be a finite family of simple lcc sheaves of A-modules on $X_{\text{ét}}$. Then there is a number N such that for every $(\{X\}, \mathcal{L})$-constructible sheaf of A-modules F on $X_{\text{ét}}$ we have*

$$\#H^i(X_{\text{ét}}, F) \leq \#F(\overline{\eta})^N,$$

for all $i \geq 0$. Here η denotes the generic point of X and $F(\overline{\eta})$ the geometric generic fiber of F.

PROOF. Take N such that

$$\#H^i(X_{\text{ét}}, L_j) \leq \#L_j(\overline{\eta})^N,$$

for all $i \geq 0$ and all $j = 1, \ldots n$. □

LEMMA 5.27. *Let A be a complete discrete valuation ring with finite residue field and field of fractions K. Let $(M_n)_{n \in \mathbb{N}}$ be an AR-ℓ-adic system of finite A-modules. Assume there exists an N such that for all $n \in \mathbb{N}$ we have*

$$\#M_n \leq \#\kappa(A)^{(n+1)N}.$$

Then

$$\dim_K K \otimes_A \mathrm{proj}\lim_n M_n \leq N.$$

PROOF. Let $(F_n)_{n \in \mathbb{N}}$ be an ℓ-adic system of finite A-modules with an AR-isomorphism $(F_n) \to (M_n)$. Then letting $F = \mathrm{proj}\lim F_n$ and $M = \mathrm{proj}\lim M_n$, we have $F \xrightarrow{\sim} M$. Therefore we have $\dim_K K \otimes_A M = \mathrm{rk}_A F$. Since (F_n) is ℓ-adic, we have for every $n \in \mathbb{N}$ that $F_n = F \otimes \Lambda_n$, and hence that $\#(\Lambda_n)^{\mathrm{rk}_A F} \leq \#F_n$. Let r be as in Lemma 1.8. Then for every n the module F_n is a direct summand of $M_{n+r} \otimes \Lambda_n$ and hence $\#F_n \leq \#M_{n+r}$. Thus we have $\#(\Lambda_n)^{\mathrm{rk}_A F} \leq \#(\Lambda_0)^{(n+r+1)N}$, which implies $(n+1)\mathrm{rk}_A F \leq (n+r+1)N$, for all $n \geq 0$. □

COROLLARY 5.28. *Let X, A and $\mathcal{L}$ be as in Lemma 5.26, but let us drop the assumption that k be algebraically closed. Let K be the quotient field of A. Then there exists an N such that for every $(\{X\}, \mathcal{L})$-constructible K-sheaf F on X we have*

$$\dim_K H^i(X, F) \leq N \mathrm{rk}\, F,$$

for all $i \geq 0$. Here $H^i(X, F)$ is used in the sense of Section 1.

PROOF. Let F be represented by the torsion free system $(F_n)_{n \in \mathbb{N}}$. Then for every $n \in \mathbb{N}$ we have $\#F_n(\overline{\eta}) = \#\kappa(A)^{(n+1)\operatorname{rk} F}$. This implies the result. $\square$

THEOREM 5.29. *Let $f : \mathfrak{X} \to \mathfrak{Y}$ be a morphism of finite type algebraic S-stacks. If $M \in \operatorname{ob} \mathbb{D}_m^+(\mathfrak{X}, \overline{\mathbb{Q}}_\ell)$ is convergent, then so is $\mathbb{R} f_* M$.*

PROOF. By the base change theorem Proposition 4.86, we may assume that $\mathfrak{Y} = \operatorname{Spec} \mathbb{F}_q$ is the spectrum of a finite field. Then by the second devissage lemma we reduce to the following two cases.

(1) $\mathfrak{X} = X$ is a smooth variety of dimension d and M is lisse.
(2) $\mathfrak{X} = BG$, where G is a smooth connected group variety.

In both cases we need to prove that $H^*(\mathfrak{X}, M)$ is convergent.

Consider Case (1). There exists a finite extension K of $\mathbb{Q}_\ell$ with ring of integers A, such that $M \in \operatorname{ob} \mathbb{D}_{(\{X\},\mathcal{L})}^+(X_{\text{ét}}, A)$, for some finite family $\mathcal{L}$ of simple lcc sheaves of A-modules on $X_{\text{ét}}$. We have

$$\sum_p \sum_n \dim \operatorname{gr}_p^w H^n(X, M) |z|^p$$

$$\leq \sum_p \sum_n \sum_{\nu=0}^{2d} \dim \operatorname{gr}_p^w H^\nu(X, h^{n-\nu} M) |z|^p$$

by the spectral sequence $H^i(X, h^j M) \Rightarrow H^{i+j}(X, M)$

$$\leq \sum_p \sum_n \sum_{\nu=0}^{2d} \sum_{p'} \dim \operatorname{gr}_p^w H^\nu(X, \operatorname{gr}_{p'}^w h^{n-\nu} M) |z|^p$$

$$\leq \sum_{\nu=0}^{2d} \sum_{p'} \sum_n \sum_p \dim \operatorname{gr}_p^w H^\nu(X, \operatorname{gr}_{p'}^w h^{n-\nu} M) |z|^{p'+\nu}$$

since we only get a contribution if $p \geq p' + \nu$,
by Corollaire 3.3.5 in [9]

$$= \sum_{\nu=0}^{2d} \sum_{p'} \sum_n \dim H^\nu(X, \operatorname{gr}_{p'}^w h^{n-\nu} M) |z|^{p'+\nu}$$

$$\leq N \sum_{\nu=0}^{2d} \sum_{p'} \sum_n \operatorname{rk} \operatorname{gr}_{p'}^w h^{n-\nu} M |z|^{p'+\nu}$$

where we choose N as in Corollary 5.28

$$= N \sum_{\nu=0}^{2d} \left(\sum_{p'} \sum_n \operatorname{rk} \operatorname{gr}_{p'}^w h^n M |z|^{p'} \right) |z|^\nu. \tag{15}$$

Since M is convergent, the term in the parentheses converges, which implies the result.

Now consider Case (2). We have

$$\sum_p \sum_n \dim \operatorname{gr}_p^w H^n(BG, M) |z|^p$$

$$\leq \sum_p \sum_n \sum_{\mu+\nu=n} \dim \operatorname{gr}_p^w H^\mu(BG, h^\nu M) |z|^p$$

$$= \sum_p \sum_n \sum_{\mu+\nu=n} \dim \operatorname{gr}_p^w \left(H^\mu(BG, Q_\ell) \otimes h^\nu M \right) |z|^p$$

$$= \sum_p \sum_n \sum_{\mu+\nu=n} \sum_{p_1+p_2=p} \dim \operatorname{gr}_{p_1}^w H^\mu(BG, Q_\ell) \dim \operatorname{gr}_{p_2}^w h^\nu M |z|^p$$

$$= \sum_p \sum_n \dim \operatorname{gr}_p^w H^n(BG, Q_\ell) |z|^p \sum_p \sum_n \dim \operatorname{gr}_p^w h^n M |z|^p. \qquad (16)$$

Thus we are reduced to the case $M = Q_\ell$.

If $F \in \operatorname{ob} \operatorname{Mod}_c(\operatorname{Spec} \mathbb{F}_q, Q_\ell)$ is pure of weight n then the i-th symmetric power $S^i F$ is pure of weight ni. So

$$\begin{aligned} w(SF)(z) &= \sum_p \sum_i \dim \operatorname{gr}_p^w S^i F \, z^p \\ &= \sum_i \dim S^i F \, z^{ni} \\ &= \left(\frac{1}{1-z^n} \right)^{\dim F}, \end{aligned}$$

which is holomorphic in the unit disc if $n > 0$. More generally, if we have pure objects $F_1, \ldots, F_n$ of $\operatorname{Mod}_c(\operatorname{Spec} \mathbb{F}_q, Q_\ell)$, then we have

$$\begin{aligned} w(S(F_1 \oplus \ldots \oplus F_n))(z) &= \prod_{i=1}^n w(SF_i)(z) \\ &= \prod_{i=1}^n \left(\frac{1}{1-z^{\operatorname{wt} F_i}} \right)^{\dim F_i}. \end{aligned}$$

This describes the structure of $w(SF)$ for every object F of $\operatorname{Mod}_c(\operatorname{Spec} \mathbb{F}_q, Q_\ell)$, since over $\operatorname{Spec} \mathbb{F}_q$ the weight filtration always splits.

Now if G is a connected group variety over $\mathbb{F}_q$, then by Theorem 5.6 there exists an object $N = N_1 \oplus \ldots \oplus N_n$ of $\operatorname{Mod}_c(\operatorname{Spec} \mathbb{F}_q, Q_\ell)$, where $N_1, \ldots, N_n$ are pure of positive weights, such that $H^*(G, Q_\ell) = \Lambda N$. (That the weights are positive follows for example from Corollaire 3.3.5 of [9].) Then by Theorem 5.6 we have $H^*(BG, Q_\ell) = SN$, which is convergent by the above considerations. $\square$

4. The Trace Formula

4.1. More about Frobenius. Let $\mathfrak{X}$ be an algebraic $\mathbb{F}_q$-stack. The stack $\mathfrak{X}$ is naturally endowed with an endomorphism $F_{\mathfrak{X}}$, the em absolute

Frobenius of $\mathfrak{X}$. For an $\mathbb{F}_q$-scheme S, interpreting objects of $\mathfrak{X}(S)$ as morphisms $S \to \mathfrak{X}$, the Frobenius $F_{\mathfrak{X}}(S)$ takes an object $x : S \to \mathfrak{X}$ to $x \circ F_S$, where $F_S : S \to S$ is the absolute Frobenius of S (which is the spectrum of the q-th power map if S is affine).

In the particular case that $S = \operatorname{Spec} \mathbb{F}_{q^n}$, for some $n > 0$, the endomorphism $F_{\mathfrak{X}}(\mathbb{F}_{q^n}) : \mathfrak{X}(\mathbb{F}_{q^n}) \to \mathfrak{X}(\mathbb{F}_{q^n})$ is endowed with a natural transformation $T_n : F_{\mathfrak{X}}(\mathbb{F}_{q^n})^n \to \operatorname{id}_{\mathfrak{X}(\mathbb{F}_{q^n})}$. Let us abbreviate notation to $T_n : F^n \to \operatorname{id}$. Now let $n, m > 0$ be such that $n \mid m$ and $m = kn$. We have the natural transformation
$$T_n \circ F^n(T_n) \circ \ldots \circ F^{(k-1)n}(T_n) : F^m \to \operatorname{id}$$
between endomorphism of $\mathfrak{X}(\mathbb{F}_{q^n})$. The compatibility property is that this natural transformation composed with $\mathfrak{X}(\mathbb{F}_{q^n}) \to \mathfrak{X}(\mathbb{F}_{q^m})$ is equal to $\mathfrak{X}(\mathbb{F}_{q^n}) \to \mathfrak{X}(\mathbb{F}_{q^m})$ composed with $T_m : F^m \to \operatorname{id}$. By abuse of notation, we may write
$$T_m = T_n \circ F^n(T_n) \circ \ldots \circ F^{(k-1)n}(T_n).$$

PROPOSITION 5.30. *If $x \in \operatorname{ob} \mathfrak{X}(\mathbb{F}_{q^n})$ is an object such that $x \cong F(x)$ in $\mathfrak{X}(\mathbb{F}_{q^n})$, then there exists an object $y \in \operatorname{ob} \mathfrak{X}(\mathbb{F}_q)$, such that $x \cong y$ in $\mathfrak{X}(\mathbb{F}_{q^m})$, for some multiple m of n.*

PROOF. Choose an isomorphism $\alpha : x \to F(x)$ in $\mathfrak{X}(\mathbb{F}_{q^n})$. Then $T_n(x) \circ F^{n-1}(\alpha) \circ \ldots \circ F(\alpha) \circ \alpha$ is an automorphism of x, in other words an element of $\operatorname{Aut}(x)(\mathbb{F}_{q^n})$. Since $\operatorname{Aut}(x)$ is a group scheme over $\mathbb{F}_{q^n}$ of finite type, the group $\operatorname{Aut}(x)(\mathbb{F}_{q^n})$ is finite. So our element $T_n(x) \circ F^{n-1}(\alpha) \circ \ldots \circ F(\alpha) \circ \alpha$ has finite order, say k, so that
$$(T_n(x) \circ F^{n-1}(\alpha) \circ \ldots \circ F(\alpha) \circ \alpha)^k = \operatorname*{id}_x.$$
We set $m = kn$. Then we have
$$(T_n(x) \circ F^{n-1}(\alpha) \circ \ldots \circ F(\alpha) \circ \alpha)^k = T_m(x) \circ F^{m-1}(\alpha) \circ \ldots \circ F(\alpha) \circ \alpha,$$
as automorphism of x in $\mathfrak{X}(\mathbb{F}_{q^m})$. this implies that
$$T_m(x) \circ F^{m-1}(\alpha) \circ \ldots \circ F(\alpha) \circ \alpha = \operatorname*{id}_x,$$
so that we have descent data for the object x of $\mathfrak{X}(\mathbb{F}_{q^m})$, giving rise to the sought after object y of $\mathfrak{X}(\mathbb{F}_q)$. □

COROLLARY 5.31. *If $\mathfrak{X}$ is an algebraic $\mathbb{F}_q$-gerbe of finite type, then $\mathfrak{X}$ is neutral.*

PROOF. We need to show that $\mathfrak{X}(\mathbb{F}_q)$ contains objects. By the Proposition, it will suffice to find an object x of $\mathfrak{X}(\mathbb{F}_{q^n})$, for some $n \geq 0$, such that $x \cong F(x)$. Since $\mathfrak{X}$ is of finite type, there exists an $n \geq 0$ and an object x of $\mathfrak{X}(\mathbb{F}_{q^n})$. Since $\mathfrak{X}$ is a gerbe, the objects x and $F(x)$ are locally isomorphic, i.e. there exists a multiple m of n such that $x \cong F(x)$ in $\mathfrak{X}(\mathbb{F}_{q^m})$. □

Let $\mathfrak{X}$ be an algebraic $\mathbb{F}_q$-gerbe of finite type. Choose an object x of $\mathfrak{X}(\mathbb{F}_q)$, and let $G = \text{Aut}(x)$. Then G is a group scheme over $\mathbb{F}_q$ of finite type and $\mathfrak{X} \cong BG$. Hence $\mathfrak{X}(\mathbb{F}_{q^n})$ is isomorphic to the groupoid of $G_{\mathbb{F}_{q^n}}$-torsors. If G is connected, by a theorem of S. Lang, these torsors are trivial. Thus $\mathfrak{X}(\mathbb{F}_{q^n})$ is a connected groupoid, for every n. We have $\#\mathfrak{X}(\mathbb{F}_{q^n}) = \frac{1}{\#G(\mathbb{F}_{q^n})}$. (Recall from Definition 3.2.1 of [**3**] that

$$\#\mathfrak{X}(\mathbb{F}_{q^n}) = \sum_{\xi \in [\mathfrak{X}(\mathbb{F}_{q^n})]} \frac{1}{\#\text{Aut}\,\xi},$$

the sum being taken over the isomorphism classes of the category $\mathfrak{X}(\mathbb{F}_{q^n})$.)

4.2. Trace of Frobenius. Let us fix an embedding $\iota : \overline{\mathbb{Q}}_\ell \to \mathbb{C}$.

Let (M, Φ) be an object of $\mathbb{D}^+_{a_{q_0}}(\text{circ}, \overline{\mathbb{Q}}_\ell)$. (See Definition 5.10.) Then if $q_0 < 1$ the series $\sum_i (-1)^i \iota(\text{tr}\,\Phi | h^i M)$ is an absolutely convergent series of complex numbers. We denote the sum by

$$\text{tr}_\iota \Phi | M = \sum_i (-1)^i \iota(\text{tr}\,\Phi | h^i M).$$

PROPOSITION 5.32. *Let*

$$\begin{array}{ccc}
 & (M'', \Phi) & \\
\swarrow & & \searrow \\
(M', \Phi) & \longrightarrow & (M, \Phi)
\end{array}$$

be a distinguished triangle in $\mathbb{D}^+_{a_{q_0}}(\text{circ}, \overline{\mathbb{Q}}_\ell)$, *where* $q_0 < 1$. *Then we have* $\text{tr}_\iota \Phi | M = \text{tr}_\iota \Phi | M' + \text{tr}_\iota \Phi | M''$.

Now let M be an object of $\mathbb{D}^+_a(\text{Spec}\,\mathbb{F}_q, \overline{\mathbb{Q}}_\ell)$. Considering M as an object of $\mathbb{D}^+_{a_{1/q}}(\text{circ}, \overline{\mathbb{Q}}_\ell)$, we denote the corresponding automorphism by Φ_q and call it the *arithmetic Frobenius* of M. Its ι-trace is denoted by

$$\text{tr}_\iota \Phi_q | M = \sum_i (-1)^i \iota(\text{tr}\,\Phi_q | h^i M).$$

It exists because $1/q < 1$. We have, for example, $\text{tr}_\iota \Phi_q | \overline{\mathbb{Q}}_\ell(n) = q^n$, for all $n \in \mathbb{Z}$.

DEFINITION 5.33. More generally, let $\mathfrak{X}$ be an algebraic $\mathbb{F}_q$-stack of finite type and $M \in \text{ob}\,\mathbb{D}^+_a(\mathfrak{X}, \overline{\mathbb{Q}}_\ell)$ a convergent $\overline{\mathbb{Q}}_\ell$-complex on $\mathfrak{X}$. We define the *local term of* M *with respect to* ι, denoted $L_\iota(M)$ by

$$L_\iota(M) = \sum_{x \in [\mathfrak{X}(\mathbb{F}_q)]} \frac{1}{\#\text{Aut}(x)}\,\text{tr}_\iota \Phi_q | \mathbb{R}x^! M,$$

where the sum is taken over all isomorphism classes of the essentially finite groupoid $\mathfrak{X}(\mathbb{F}_q)$.

In the following considerations ι is fixed, so that we will drop it from our notation.

LEMMA 5.34. *We have*
$$L(M) = \sum_{x \in [\mathfrak{X}(\mathbb{F}_q)]} \frac{1}{\#\operatorname{Aut}(x)} L(\mathbb{R}x^! M).$$

PROOF. This follows immediately from the fact that for $\mathfrak{X} = \operatorname{Spec}\mathbb{F}_q$ we have $L(M) = \operatorname{tr}\Phi_q | M$. □

LEMMA 5.35. *If $i : \mathfrak{Z} \to \mathfrak{X}$ is a closed immersion with complement $j : \mathfrak{U} \to \mathfrak{X}$ we have*
$$L(M) = L(\mathbb{R}i^! M) + L(j^* M).$$

PROOF. This is a formal consequence of Lemma 5.34 using the fact that $[\mathfrak{X}(\mathbb{F}_q)]$ is the disjoint union of $[\mathfrak{U}(\mathbb{F}_q)]$ and $[\mathfrak{Z}(\mathbb{F}_q)]$. □

LEMMA 5.36. *If*
$$\begin{array}{ccc} & M'' & \\ \swarrow & & \nwarrow \\ M' & \longrightarrow & M \end{array}$$
is a distinguished triangle in $\mathbb{D}_a^+(\mathfrak{X}, Q_\ell)$ then we have $L(M) = L(M') + L(M'')$.

PROOF. This follows immediately from Proposition 5.32. □

LEMMA 5.37. *Let $\mathfrak{X}$ be smooth of dimension $d \in \mathbb{Z}$ over $\mathbb{F}_q$, and assume that $M \in \operatorname{ob}\mathbb{D}_a^+(\mathfrak{X}, Q_\ell)$ is lisse. Then we have*
$$q^d L(M) = \sum_{x \in [\mathfrak{X}(\mathbb{F}_q)]} \frac{1}{\#\operatorname{Aut}(x)} L(x^* M).$$

PROOF. This follows from Proposition 4.87. □

THEOREM 5.38. *Let $f : \mathfrak{X} \to \mathfrak{Y}$ be a morphism of finite type algebraic $\mathbb{F}_q$-stacks. Then for every convergent Q_ℓ-complex $M \in \operatorname{ob}\mathbb{D}_a^+(\mathfrak{X}, Q_\ell)$ we have*
$$L(\mathbb{R}f_* M) = L(M).$$

PROOF. We have
$$\begin{aligned} L(\mathbb{R}f_* M) &= \sum_{y \in [\mathfrak{Y}(\mathbb{F}_q)]} \frac{1}{\#\operatorname{Aut}(y)} L(\mathbb{R}y^! \mathbb{R}f_* M) \\ &= \sum_{y \in [\mathfrak{Y}(\mathbb{F}_q)]} \frac{1}{\#\operatorname{Aut}(y)} L(\mathbb{R}f_{y*} \mathbb{R}u_y^! M), \end{aligned}$$
where the notations are taken from the 2-cartesian diagram
$$\begin{array}{ccc} \mathfrak{X}_y & \xrightarrow{f_y} & \operatorname{Spec}\mathbb{F}_q \\ u_y \downarrow & & \downarrow y \\ \mathfrak{X} & \xrightarrow{f} & \mathfrak{Y}. \end{array}$$

Assuming for the moment that the theorem holds for f_y, we get

$$\begin{aligned}
L(\mathbb{R}f_*M) &= \sum_{y \in [\mathfrak{Y}(\mathbb{F}_q)]} \frac{1}{\#\operatorname{Aut}(y)} L(\mathbb{R}u_y^! M) \\
&= \sum_{y \in [\mathfrak{Y}(\mathbb{F}_q)]} \frac{1}{\#\operatorname{Aut}(y)} \sum_{x \in [\mathfrak{X}_y(\mathbb{F}_q)]} \frac{1}{\#\operatorname{Aut}(x)} L(\mathbb{R}x^! \mathbb{R}u_y^! M).
\end{aligned}$$

Now the category $\mathfrak{X}_y(\mathbb{F}_q)$ has objects (x, α), where x is an object of $\mathfrak{X}(\mathbb{F}_q)$ and $\alpha : f(x) \to y$ is an isomorphism in $\mathfrak{Y}(\mathbb{F}_q)$. A morphism $\phi : (x, \alpha) \to (x', \alpha')$ is a morphism $\phi : x \to x'$ in $\mathfrak{X}(\mathbb{F}_q)$ such that $\alpha' \circ f(\phi) = \alpha$. So a more accurate way of writing our equation is

$$L(\mathbb{R}f_*M) = \sum_{y \in [\mathfrak{Y}(\mathbb{F}_q)]} \frac{1}{\#\operatorname{Aut}(y)} \sum_{(x,\alpha) \in [\mathfrak{X}_y(\mathbb{F}_q)]} \frac{1}{\#\operatorname{Aut}(x,\alpha)} L(\mathbb{R}x^! M).$$

Now we have an action of $\operatorname{Aut}(y)$ on $[\mathfrak{X}_y(\mathbb{F}_q)]$, given by $\beta \cdot (x, \alpha) = (x, \beta\alpha)$, for $\beta \in \operatorname{Aut}(y)$ and $(x, \alpha) \in [\mathfrak{X}_y(\mathbb{F}_q)]$. This action has the property that (x, α) and (x', α') are in the same orbit, if and only if $x \cong x'$ in $\mathfrak{X}(\mathbb{F}_q)$. The isotropy group of an element (x, α) of $[\mathfrak{X}_y(\mathbb{F}_q)]$ is the image of $\operatorname{Aut}(x)$ in $\operatorname{Aut}(y)$, hence isomorphic to $\operatorname{Aut}(x)/\operatorname{Aut}(x, \alpha)$. So we have

$$\begin{aligned}
&L(\mathbb{R}f_*M) \\
&= \sum_{y \in [\mathfrak{Y}(\mathbb{F}_q)]} \frac{1}{\#\operatorname{Aut}(y)} \sum_{\substack{x \in [\mathfrak{X}(\mathbb{F}_q)] \\ x \mapsto y}} \left(\sum_{\substack{(x,\alpha) \in [\mathfrak{X}_y(\mathbb{F}_q)] \\ (x,\alpha) \mapsto x}} \frac{1}{\#\operatorname{Aut}(x,\alpha)} \right) L(\mathbb{R}x^! M) \\
&= \sum_{y \in [\mathfrak{Y}(\mathbb{F}_q)]} \frac{1}{\#\operatorname{Aut}(y)} \sum_{\substack{x \in [\mathfrak{X}(\mathbb{F}_q)] \\ x \mapsto y}} \frac{1}{\#\operatorname{Aut}(x)} \left(\sum_{\substack{(x,\alpha) \in [\mathfrak{X}_y(\mathbb{F}_q)] \\ (x,\alpha) \mapsto x}} \frac{\#\operatorname{Aut}(x)}{\#\operatorname{Aut}(x,\alpha)} \right) L(\mathbb{R}x^! M) \\
&= \sum_{y \in [\mathfrak{Y}(\mathbb{F}_q)]} \frac{1}{\#\operatorname{Aut}(y)} \sum_{\substack{x \in [\mathfrak{X}(\mathbb{F}_q)] \\ x \mapsto y}} \frac{1}{\#\operatorname{Aut}(x)} \#\operatorname{Aut}(y) L(\mathbb{R}x^! M) \\
&= \sum_{y \in [\mathfrak{Y}(\mathbb{F}_q)]} \sum_{\substack{x \in [\mathfrak{X}(\mathbb{F}_q)] \\ x \mapsto y}} \frac{1}{\#\operatorname{Aut}(x)} L(\mathbb{R}x^! M) \\
&= \sum_{x \in [\mathfrak{X}(\mathbb{F}_q)]} \frac{1}{\#\operatorname{Aut}(x)} L(\mathbb{R}x^! M) \\
&= L(M),
\end{aligned}$$

which is what we wanted to prove. It remains to prove the theorem for the various f_y. So without loss of generality we may assume that $\mathfrak{Y} = \operatorname{Spec} \mathbb{F}_q$.

But let us still consider the general case. Assume that $\mathfrak{X}$ is the disjoint union of a closed substack $i : \mathfrak{Z} \to \mathfrak{X}$ and an open substack $j : \mathfrak{U} \to \mathfrak{X}$.

4. THE TRACE FORMULA

Assuming the theorem to hold for $f \circ i$ and $f \circ j$ we get

$$\begin{aligned} L(\mathbb{R}f_*M) &= L(\mathbb{R}(f \circ i)_*\mathbb{R}i^!M) + L(\mathbb{R}(f \circ j)_*j^*M) \\ &= L(\mathbb{R}i^!M) + L(j^*M) \\ &= L(M). \end{aligned}$$

Thus we may pass to the strata of any stratification of $\mathfrak{X}$ in proving our theorem. So we have reduced to the case $\mathfrak{Y} = \operatorname{Spec} \mathbb{F}_q$, $\mathfrak{X}$ is either a smooth $\mathbb{F}_q$-variety or an $\mathbb{F}_q$-gerbe, and $M \in \operatorname{ob} \mathbb{D}_a^+(\mathfrak{X}, \overline{\mathbb{Q}}_\ell)$ is lisse.

From the inequalities (15) and (16) it follows that $\sum_n L(\mathbb{R}f_*h^nM)$ is absolutely convergent and that $\lim_{n\to\infty} L(\mathbb{R}f_*\tau_{\geq n}M) = 0$. By the distinguished triangle

$$\begin{array}{ccc} & \tau_{\geq n+1}M & \\ \swarrow & & \nwarrow \\ h^n(M) & \longrightarrow & \tau_{\geq n}M \end{array}$$

this implies that

$$\begin{aligned} L(\mathbb{R}f_*M) &= \sum_n L(\mathbb{R}f_*h^nM) \\ &= \sum_n L(h^nM) \\ &= L(M), \end{aligned}$$

if we assume the truth of the theorem for lisse $\overline{\mathbb{Q}}_\ell$-sheaves. So without loss of generality M is a lisse $\overline{\mathbb{Q}}_\ell$-sheaf on $\mathfrak{X}$.

Let us first consider the case that $\mathfrak{X} = X$ is a smooth $\mathbb{F}_q$-variety of dimension d. We need to show that

$$q^d \sum_{i=1}^{2d} (-1)^i \operatorname{tr} \Phi_q | H^i(\overline{X}_{\text{ét}}, M) = \sum_{x \in X(\mathbb{F}_q)} \operatorname{tr} \Phi_q | x^*M.$$

Now from Poincaré duality we have

$$\begin{aligned} q^d \sum_{i=1}^{2d} (-1)^i \operatorname{tr} \Phi_q | H^i(\overline{X}_{\text{ét}}, M) &= \sum_{i=1}^{2d} (-1)^i \operatorname{tr} \Phi_q | H^i(\overline{X}_{\text{ét}}, M(d)) \\ &= \sum_{i=1}^{2d} (-1)^i \operatorname{tr} \Phi_q^\vee | H^i(\overline{X}_{\text{ét}}, M(d))^\vee \\ &= \sum_{i=1}^{2d} (-1)^i \operatorname{tr} F_q | H_c^{2d-i}(\overline{X}_{\text{ét}}, M^\vee) \\ &= \sum_{i=1}^{2d} (-1)^i \operatorname{tr} F_q | H_c^i(\overline{X}_{\text{ét}}, M^\vee), \end{aligned}$$

where F_q is the geometric Frobenius. The 'usual' Lefschetz Trace Formula for the geometric Frobenius on cohomology with compact supports reads

$$\sum_{i=1}^{2d}(-1)^i \operatorname{tr} F_q|H_c^i(\overline{X}_{\text{ét}}, M^\vee) = \sum_{x\in X(\mathbb{F}_q)} \operatorname{tr} F_q|x^*M^\vee.$$

But clearly, $\operatorname{tr} F_q|x^*M^\vee = \operatorname{tr} \Phi_q|x^*M$. This finishes the proof in the case that $\mathfrak{X}$ is a variety.

So now assume that $\mathfrak{X}$ is a gerbe. By Corollary 5.31 we may assume that $\mathfrak{X} = BG$, where G is an algebraic group over $\mathbb{F}_q$. As in the second devissage lemma, we may reduce to the case that G is smooth and connected. In this case we have $\operatorname{Mod}_m(BG, Q_\ell) = \operatorname{Mod}_m(\operatorname{Spec} \mathbb{F}_q, Q_\ell)$, so we may assume that $M = f^*N$, for some mixed Q_ℓ-vector space N. Then we have

$$\begin{aligned} L(\mathbb{R}f_*M) &= L(\mathbb{R}f_*f^*N) \\ &= L(\mathbb{R}f_*Q_\ell \otimes N) \\ &= L(\mathbb{R}f_*Q_\ell)L(N) \\ &= L(Q_{\ell\mathfrak{X}})L(N), \end{aligned}$$

assuming the theorem holds for $Q_{\ell\mathfrak{X}}$. Now by Lemma 5.37 we have

$$\begin{aligned} L(M) &= L(f^*N) \\ &= \frac{1}{q^{\dim \mathfrak{X}}} \sum_{x\in[\mathfrak{X}(\mathbb{F}_q)]} \frac{1}{\operatorname{Aut}(x)} L(x^*f^*N) \\ &= \frac{1}{q^{\dim \mathfrak{X}}} \sum_{x\in[\mathfrak{X}(\mathbb{F}_q)]} \frac{1}{\operatorname{Aut}(x)} L(N) \\ &= L(Q_{\ell\mathfrak{X}})L(N). \end{aligned}$$

So we are now reduced to the case that $M = Q_\ell$. We need to prove that

$$\sum_{i=0}^{\infty}(-1)^i \operatorname{tr} \Phi_q|H^i(BG, Q_\ell) = \frac{1}{q^{\dim BG}} \sum_{x\in[BG(\mathbb{F}_q)]} \frac{1}{\#\operatorname{Aut}(x)},$$

or in other words that

$$\sum_{i=0}^{\infty}(-1)^i \operatorname{tr} \Phi_q|H^i(BG, Q_\ell) = q^{\dim G} \frac{1}{\#G(\mathbb{F}_q)}.$$

So let us finally assume that G is a smooth group variety over $\mathbb{F}_q$. Let $N = \bigoplus_i N^i$ be the transgressive submodule of $\bigoplus_i H^i(G, Q_\ell)$. Then Φ_q acts on N. Let $\lambda_1, \ldots, \lambda_r$ be the eigenvalues of Φ_q on N. Then we have

$$\operatorname{tr} \Phi_q|\Lambda N = \prod_{j=1}^r (1-\lambda_j),$$

since N is concentrated in odd degrees. Similarly,

$$\operatorname{tr} \Phi_q | S(N[-1]) = \prod_{j=1}^{r} \frac{1}{1-\lambda_j}.$$

Now by Theorem 5.6 we have

$$\begin{aligned}
\operatorname{tr} \Phi_q | \Lambda N &= \operatorname{tr} \Phi_q | H^*(G, Q_\ell) \\
&= \frac{1}{q^{\dim G}} \#G(\mathbb{F}_q),
\end{aligned}$$

by the earlier result about varieties. On the other hand we have

$$\begin{aligned}
\operatorname{tr} \Phi_q | H^*(BG, Q_\ell) &= \operatorname{tr} \Phi_q | S(N[-1]) \\
&= \prod_{j=1}^{r} \frac{1}{1-\lambda_j} \\
&= \left(\prod_{j=1}^{r} (1-\lambda_j) \right)^{-1} \\
&= (\operatorname{tr} \Phi_q | \Lambda N)^{-1} \\
&= q^{\dim G} \frac{1}{\#G(\mathbb{F}_q)},
\end{aligned}$$

which is what we needed to prove. $\square$

COROLLARY 5.39. *Let $\mathfrak{X}$ be a smooth algebraic $\mathbb{F}_q$-stack of finite type of dimension $d \in \mathbb{Z}$. Let $M \in \operatorname{ob} \mathbb{D}_a^+(\mathfrak{X}, Q_\ell)$ be a lisse convergent Q_ℓ-complex on $\mathfrak{X}$ (for example a bounded mixed lisse Q_ℓ-complex). Then we have*

$$q^{\dim \mathfrak{X}} \operatorname{tr} \Phi_q | H^*(\mathfrak{X}, M) = \sum_{x \in [\mathfrak{X}(\mathbb{F}_q)]} \frac{1}{\#\operatorname{Aut}(x)} \operatorname{tr} \Phi_q | x^* M.$$

For example, for $M = \mathbb{Q}_\ell$, we get

$$q^{\dim \mathfrak{X}} \operatorname{tr} \Phi_q | H^*(\mathfrak{X}, \mathbb{Q}_\ell) = \#\mathfrak{X}(\mathbb{F}_q).$$

This is the result we conjectured in [**3**].

4.3. An Example. Let us apply our trace formula to the algebraic stack $\mathfrak{M}_1$, of curves of genus one.

PROPOSITION 5.40. *Let p be a prime number. Then we have*

$$\sum_k \frac{1}{p^{k+1}} \operatorname{tr} T_p | \mathcal{S}_{k+2} = 1 - \frac{1}{p^3 - p} - \sum_{E/\mathbb{F}_p} \frac{1}{\#\operatorname{Aut}(E) \# E(\mathbb{F}_p)}.$$

Here T_p is the pth Hecke operator on the space of cusp forms $\mathcal{S}_{k+2}$ of weight $k+2$. The sum on the right hand side extends over all isomorphism classes of elliptic curves over $\mathbb{F}_p$.

PROOF. Let $\mathfrak{M}_1$ be the algebraic $\mathbb{Z}$-stack of curves of genus $g = 1$ and $\mathfrak{M}_{1,1}$ the algebraic $\mathbb{Z}$-stack of elliptic curves. Via the obvious morphism $\mathfrak{M}_{1,1} \to \mathfrak{M}_1$ we may think of $\mathfrak{M}_{1,1}$ as the universal curve of genus one. Via the morphism $\pi : \mathfrak{M}_1 \to \mathfrak{M}_{1,1}$, associating with a curve of genus one its Jacobian, we may think of $\mathfrak{M}_1$ as $B(\mathfrak{E}/\mathfrak{M}_{1,1})$, the classifying stack of the universal elliptic curve $f : \mathfrak{E} \to \mathfrak{M}_{1,1}$. By Corollary 5.39 we have

$$\operatorname{tr} \Phi_p | H^*(\mathfrak{M}_1, \mathbb{Q}_\ell) = \#\mathfrak{M}_1(\mathbb{F}_p) = \sum_{E/\mathbb{F}_p} \frac{1}{\#\operatorname{Aut}(E) \# E(\mathbb{F}_p)},$$

noting that $\dim \mathfrak{M}_1 = 3g - 3 = 0$. We will show that

$$\operatorname{tr} \Phi_p | H^*(\mathfrak{M}_1, \mathbb{Q}_\ell) = 1 - \frac{1}{p^3 - p} - \sum_k \frac{1}{p^{k+1}} \operatorname{tr} T_p | S_{k+2}.$$

Let us abbreviate the local system $R^1 f_* \mathbb{Q}_\ell$ on $\mathfrak{M}_{1,1}$ by $H^1(\mathfrak{E})$. We claim that

$$\begin{aligned} H^0(\mathfrak{M}_1, \mathbb{Q}_\ell) &= \mathbb{Q}_\ell \\ H^{2k+1}(\mathfrak{M}_1, \mathbb{Q}_\ell) &= H^1(\mathfrak{M}_{1,1}, S^k H^1(\mathfrak{E})), \quad \text{for all } k \geq 0 \\ H^i(\mathfrak{M}_1, \mathbb{Q}_\ell) &= 0, \quad \text{if } i \text{ is even, } i \geq 2. \end{aligned} \quad (17)$$

To see this, note that by Remark 5.7 we may write the Leray spectral sequence of π as

$$H^i(\mathfrak{M}_{1,1}, S^k H^1(\mathfrak{E})) \Longrightarrow H^{2k+i}(\mathfrak{M}_1, \mathbb{Q}_\ell). \quad (18)$$

To analyze the E_2-term in this spectral sequence, note that

$$H^i(\mathfrak{M}_{1,1}^{\mathrm{an}}, S^k H^1(\mathfrak{E})) = H^i(SL_2\mathbb{Z}, M_k),$$

where M_k is the natural representation of $SL_2\mathbb{Z}$ on the homogeneous polynomials in two variables of degree k. This follows from the fact that $\mathfrak{M}_{1,1}^{\mathrm{an}} = SL_2\mathbb{Z} \backslash \mathbb{H}$, where $\mathbb{H}$ is the complex upper half plane. Now it is well-known (and elementary) that

$$\begin{aligned} \dim H^0(SL_2\mathbb{Z}, M_0) &= 1 \\ \dim H^1(SL_2\mathbb{Z}, M_k) &= k - 1 - 2[\tfrac{k}{4}] - 2[\tfrac{k}{6}], \quad \text{if } k \text{ is even, } k \geq 2 \\ H^i(SL_2\mathbb{Z}, M_k) &= 0, \quad \text{for all other } i \text{ and } k. \end{aligned}$$

This gives the dimensions of the components of the E_2-term in the spectral sequence (18). We deduce the above Claim (17), and that

$$\begin{aligned} \dim H^{2k+1}(\mathfrak{M}_1, \mathbb{Q}_\ell) &= k - 1 - 2[\tfrac{k}{4}] - 2[\tfrac{k}{6}], \quad \text{if } k \text{ is even, } k \geq 2 \\ H^{2k+1}(\mathfrak{M}_1, \mathbb{Q}_\ell) &= 0, \quad \text{otherwise.} \end{aligned}$$

So $H^i(\mathfrak{M}_1, \mathbb{Q}_\ell) \neq 0$ if and only if $i \equiv 1 \mod (4)$, except for $i = 0, 1$.

Now let $\mathfrak{T}_{p,1}$ be the algebraic $\mathbb{Z}$-stack of isogenies of elliptic curves of degree p. The stack $\mathfrak{T}_{p,1}$ comes with a pair of morphisms $\mathfrak{T}_{p,1} \xrightarrow[t]{s} \mathfrak{M}_{1,1}$ and is called the Hecke correspondence. We get an induced endomorphism T_p of $H^1(\mathfrak{M}_{1,1_{\mathbb{F}_p}}, S^k H^1(\mathfrak{E}))$ and and induced endomorphism T_p of $H^1(SL_2\mathbb{Z}, M_k)$. We will relate $T_p | H^1(\mathfrak{M}_{1,1_{\mathbb{F}_p}}, S^k H^1(\mathfrak{E}))$ to the arithmetic Frobenius Φ_p and

4. THE TRACE FORMULA

$T_p|H^1(SL_2\mathbb{Z}, M_k)$ to the Hecke operator $T_p|\mathcal{S}_{k+1}$ on the cusp forms of weight $k+2$.

In characteristic p, the isogenies of degree p of elliptic curves are essentially given by the Frobenius F and its dual V, the Verschiebung. This implies that on $H^1(\mathfrak{M}_{1,1_{\mathbb{F}_p}}, S^k H^1(\mathfrak{E}))$ we have $T_p = F+V$. It is easy to check that $FV = VF = p^{k+1}$. This implies $V = p^{k+1}\Phi_p$ and $T_p = \Phi_p^{-1} + p^{k+1}\Phi_p$. (See [7] for the details.)

On the other hand, it is a theorem of Shimura, that

$$H^1(SL_2\mathbb{Z}, M_k \otimes \mathbb{C}) = \mathcal{S}_{k+2} \oplus \overline{\mathcal{S}}_{k+2} \oplus \mathcal{E}_{k+2},$$

where $\mathcal{E}_{k+2} = \mathbb{C}E_{\frac{k+2}{2}}$ and $E_{\frac{k+2}{2}}$ is the Eisenstein series. The Hecke operator corresponds to $T_p \oplus \overline{T}_p \oplus T_p$ under this isomorphism.

Now let $\widetilde{H}^1(\mathfrak{M}_{1,1}, S^k H^1(\mathfrak{E}))$ be the image of $H_c^1(\mathfrak{M}_{1,1}, S^k H^1(\mathfrak{E}))$ in $H^1(\mathfrak{M}_{1,1}, S^k H^1(\mathfrak{E}))$. Then $\widetilde{H}^1(\mathfrak{M}_{1,1}, S^k H^1(\mathfrak{E}))$ corresponds to $\mathcal{S}_{k+2} \oplus \overline{\mathcal{S}}_{k+2}$. The vector space $\widetilde{H}^1(\mathfrak{M}_{1,1}, S^k H^1(\mathfrak{E}))$ is endowed with an inner product making F and V adjoints of each other. Thus we have

$$\begin{aligned}
\operatorname{tr} \Phi_p | \widetilde{H}^1 &= \frac{1}{p^{k+1}} \operatorname{tr} V | \widetilde{H}^1 \\
&= \frac{1}{p^{k+1}} \frac{1}{2}(\operatorname{tr} V + \operatorname{tr} F) | \widetilde{H}^1 \\
&= \frac{1}{p^{k+1}} \frac{1}{2} \operatorname{tr} T_p | \widetilde{H}^1 \\
&= \frac{1}{p^{k+1}} \frac{1}{2}(\operatorname{tr} T_p | \mathcal{S} + \operatorname{tr} \overline{T}_p | \overline{\mathcal{S}}) \\
&= \frac{1}{p^{k+1}} \operatorname{tr} T_p | \mathcal{S}_{k+2}.
\end{aligned}$$

On the other hand, since $\mathcal{E}_{k+2}$ is one-dimensional, T_p and Φ_p act as scalars on it. We have $T_p = \sigma_{k+1}(p) = \sum_{d|p} d^{k+1} = 1 + p^{k+1}$ and $T_p = \Phi_p^{-1} + p^{k+1}\Phi_p$, which implies $\Phi_p^{-1} + p^{k+1}\Phi_p = 1 + p^{k+1}$ and hence $\Phi_p = 1$ or $\Phi_p = \frac{1}{p^{k+1}}$. But by [9] $\Phi_p = 1$ is impossible, and thus $\Phi_p|\mathcal{E}_{k+2} = \frac{1}{p^{k+1}}$.

So, finally, we have

$$\begin{aligned}
\operatorname{tr} \Phi_p | H^*(\mathfrak{M}_1, \mathbb{Q}_\ell) &= 1 - \sum_{\substack{k=2 \\ 2|k}}^{\infty} \operatorname{tr} \Phi_p | H^1(\mathfrak{M}_{1,1}, S^k H^1(\mathfrak{E})) \\
&= 1 - \sum_k \operatorname{tr} \Phi_p | \mathcal{E}_{k+2} - \sum_k \operatorname{tr} \Phi_p | \widetilde{H}^1(\mathfrak{M}_{1,1}, S^k H^1(\mathfrak{E})) \\
&= 1 - \sum_k \frac{1}{p^{k+1}} - \sum_k \frac{1}{p^{k+1}} \operatorname{tr} T_p | \mathcal{S}_{k+2} \\
&= 1 - \frac{1}{p^3 - p} - \sum_k \frac{1}{p^{k+1}} \operatorname{tr} T_p | \mathcal{S}_{k+2},
\end{aligned}$$

which is what we wanted to prove. □

EXAMPLE 5.41. For the first three primes we get

$$\sum_k \frac{1}{2^{k+1}} \operatorname{tr} T_2 | \mathcal{S}_{k+2} = -\frac{1}{120},$$

$$\sum_k \frac{1}{3^{k+1}} \operatorname{tr} T_3 | \mathcal{S}_{k+2} = \frac{1}{840},$$

$$\sum_k \frac{1}{5^{k+1}} \operatorname{tr} T_5 | \mathcal{S}_{k+2} = \frac{1}{10080}.$$

Bibliography

[1] M. Artin. Versal deformations and algebraic stacks. *Inventiones mathematicae*, 27:165–189, 1974.

[2] M. Artin, A. Grothendieck, and J. L. Verdier. *Théorie des Topos et Cohomologie Etale des Schémas*, **SGA4**. Lecture Notes in Mathematics Nos. 269, 270, 305. Springer, Berlin, Heidelberg, New York, 1972, 73.

[3] K. Behrend. The Lefschetz trace formula for algebraic stacks. *Inventiones mathematicae*, 112:127–149, 1993.

[4] A. A. Beilinson, J. Bernstein, and P. Deligne. Faisceaux pervers. In *Analyse et Topologie sur les Espaces Singuliers I, Astérisque 100*. Société Mathématique de France, 1982.

[5] J. Bernstein and V. Lunts. *Equivariant Sheaves and Functors*. Lecture Notes in Mathematics No. 1578. Springer, Berlin, Heidelberg, 1994.

[6] A. Borel. Sur la cohomologie des éspaces fibrés principaux et des éspaces homogènes de groupes de Lie compacts. *Annals of Mathematics*, 57:115–207, 1953.

[7] P. Deligne. Travaux de Shimura. *Séminaire Bourbaki*, 23e année(389), 1971.

[8] P. Deligne. *Cohomologie Etale*, **SGA4$\frac{1}{2}$**. Lecture Notes in Mathematics No. 569. Springer, Berlin, Heidelberg, New York, 1977.

[9] P. Deligne. La conjecture de Weil. II. *Publications Mathématiques, Institut des Hautes Études Scientifiques*, 52:137–252, 1980.

[10] M. Demazure and A. Grothendieck. *Schémas en Groupes* **SGA3**. Lecture Notes in Mathematics Nos. 151, 152, 153. Springer, Berlin, Heidelberg, New York, 1970.

[11] T. Ekedahl. On the adic formalism. In *Grothendieck Festschrift*. Birkhäuser, 1990.

[12] A. Grothendieck et. al. *Revêtements Étales et Groupe Fondamental* **SGA1**. Lecture Notes in Mathematics No. 224. Springer, Berlin, Heidelberg, New York, 1971.

[13] A. Grothendieck et. al. *Cohomologie ℓ-adique et Fonctions L*, **SGA5**. Lecture Notes in Mathematics No. 589. Springer, Berlin, Heidelberg, New York, 1977.

[14] P.-P. Grivel. Categories dérivées et foncteurs dérivés. In A. Borel, editor, *Algebraic D-Modules, Perspectives in Mathematics, Vol. 2*. Academic Press, Orlando, 1999.

[15] A. Grothendieck. *Elements de Géométrie Algébrique IV: Etude Locale des Schémas et des Morphismes de Schémas*, **EGA4**. Publications Mathématiques Nos. 20, 24, 28, 32. Institut des Hautes Études Scientifiques, Bois-Marie Bures-sur-Yvette, 1964, 65, 66, 67.

[16] G. Laumon. *Champs Algébriques*. Prépublications No. 88–33. Université de Paris-Sud, Mathématiques, Orsay, 1988.

[17] J. S. Milne. *Arithmetic Duality Theorems*. Perspectives in Mathematics Vol. 1. Academic Press, Orlando, Florida, 1986.

Editorial Information

To be published in the *Memoirs*, a paper must be correct, new, nontrivial, and significant. Further, it must be well written and of interest to a substantial number of mathematicians. Piecemeal results, such as an inconclusive step toward an unproved major theorem or a minor variation on a known result, are in general not acceptable for publication. Papers appearing in *Memoirs* are generally longer than those appearing in *Transactions*, which shares the same editorial committee.

As of February 1, 2003, the backlog for this journal was approximately 3 volumes. This estimate is the result of dividing the number of manuscripts for this journal in the Providence office that have not yet gone to the printer on the above date by the average number of monographs per volume over the previous twelve months, reduced by the number of volumes published in four months (the time necessary for preparing a volume for the printer). (There are 6 volumes per year, each containing at least 4 numbers.)

A Consent to Publish and Copyright Agreement is required before a paper will be published in the *Memoirs*. After a paper is accepted for publication, the Providence office will send a Consent to Publish and Copyright Agreement to all authors of the paper. By submitting a paper to the *Memoirs*, authors certify that the results have not been submitted to nor are they under consideration for publication by another journal, conference proceedings, or similar publication.

Information for Authors

Memoirs are printed from camera copy fully prepared by the author. This means that the finished book will look exactly like the copy submitted.

The paper must contain a *descriptive title* and an *abstract* that summarizes the article in language suitable for workers in the general field (algebra, analysis, etc.). The *descriptive title* should be short, but informative; useless or vague phrases such as "some remarks about" or "concerning" should be avoided. The *abstract* should be at least one complete sentence, and at most 300 words. Included with the footnotes to the paper should be the 2000 *Mathematics Subject Classification* representing the primary and secondary subjects of the article. The classifications are accessible from www.ams.org/msc/. The list of classifications is also available in print starting with the 1999 annual index of *Mathematical Reviews*. The Mathematics Subject Classification footnote may be followed by a list of *key words and phrases* describing the subject matter of the article and taken from it. Journal abbreviations used in bibliographies are listed in the latest *Mathematical Reviews* annual index. The series abbreviations are also accessible from www.ams.org/publications/. To help in preparing and verifying references, the AMS offers MR Lookup, a Reference Tool for Linking, at www.ams.org/mrlookup/. When the manuscript is submitted, authors should supply the editor with electronic addresses if available. These will be printed after the postal address at the end of the article.

Electronically prepared manuscripts. The AMS encourages electronically prepared manuscripts, with a strong preference for $\mathcal{A}_{\mathcal{M}}\mathcal{S}$-LaTeX. To this end, the Society has prepared $\mathcal{A}_{\mathcal{M}}\mathcal{S}$-LaTeX author packages for each AMS publication. Author packages include instructions for preparing electronic manuscripts, the *AMS Author Handbook*, samples, and a style file that generates the particular design specifications of that publication series. Though $\mathcal{A}_{\mathcal{M}}\mathcal{S}$-LaTeX is the highly preferred format of TeX, author packages are also available in $\mathcal{A}_{\mathcal{M}}\mathcal{S}$-TeX.

Authors may retrieve an author package from e-MATH starting from `www.ams.org/tex/` or via FTP to `ftp.ams.org` (login as `anonymous`, enter username as password, and type `cd pub/author-info`). The *AMS Author Handbook* and the *Instruction Manual* are available in PDF format following the author packages link from `www.ams.org/tex/`. The author package can be obtained free of charge by sending email to `pub@ams.org` (Internet) or from the Publication Division, American Mathematical Society, P.O. Box 6248, Providence, RI 02940-6248. When requesting an author package, please specify $\mathcal{AMS}$-LaTeX or $\mathcal{AMS}$-TeX, Macintosh or IBM (3.5) format, and the publication in which your paper will appear. Please be sure to include your complete mailing address.

Sending electronic files. After acceptance, the source file(s) should be sent to the Providence office (this includes any TeX source file, any graphics files, and the DVI or PostScript file).

Before sending the source file, be sure you have proofread your paper carefully. The files you send must be the EXACT files used to generate the proof copy that was accepted for publication. For all publications, authors are required to send a printed copy of their paper, which exactly matches the copy approved for publication, along with any graphics that will appear in the paper.

TeX files may be submitted by email, FTP, or on diskette. The DVI file(s) and PostScript files should be submitted only by FTP or on diskette unless they are encoded properly to submit through email. (DVI files are binary and PostScript files tend to be very large.)

Electronically prepared manuscripts can be sent via email to `pub-submit@ams.org` (Internet). The subject line of the message should include the publication code to identify it as a Memoir. TeX source files, DVI files, and PostScript files can be transferred over the Internet by FTP to the Internet node `e-math.ams.org` (130.44.1.100).

Electronic graphics. Comprehensive instructions on preparing graphics are available at `www.ams.org/jourhtml/graphics.html`. A few of the major requirements are given here.

Submit files for graphics as EPS (Encapsulated PostScript) files. This includes graphics originated via a graphics application as well as scanned photographs or other computer-generated images. If this is not possible, TIFF files are acceptable as long as they can be opened in Adobe Photoshop or Illustrator. No matter what method was used to produce the graphic, it is necessary to provide a paper copy to the AMS.

Authors using graphics packages for the creation of electronic art should also avoid the use of any lines thinner than 0.5 points in width. Many graphics packages allow the user to specify a "hairline" for a very thin line. Hairlines often look acceptable when proofed on a typical laser printer. However, when produced on a high-resolution laser imagesetter, hairlines become nearly invisible and will be lost entirely in the final printing process.

Screens should be set to values between 15% and 85%. Screens which fall outside of this range are too light or too dark to print correctly. Variations of screens within a graphic should be no less than 10%.

Inquiries. Any inquiries concerning a paper that has been accepted for publication should be sent directly to the Electronic Prepress Department, American Mathematical Society, P. O. Box 6248, Providence, RI 02940-6248.

Editors

This journal is designed particularly for long research papers, normally at least 80 pages in length, and groups of cognate papers in pure and applied mathematics. Papers intended for publication in the *Memoirs* should be addressed to one of the following editors. In principle the Memoirs welcomes electronic submissions, and some of the editors, those whose names appear below with an asterisk (*), have indicated that they prefer them. However, editors reserve the right to request hard copies after papers have been submitted electronically. Authors are advised to make preliminary email inquiries to editors about whether they are likely to be able to handle submissions in a particular electronic form.

Algebra to KAREN E. SMITH, Department of Mathematics, University of Michigan, 525 University, Suite 2832, Ann Arbor, MI 48109-1109; email: `kesmith@lsa.umich.edu`

Algebraic geometry to DAN ABRAMOVICH, Department of Mathematics, Boston University, 111 Cummington Street, Boston, MA 02215; e-mail: `abrmovic@bu.edu`

Algebraic topology and cohomology of groups to STEWART PRIDDY, Department of Mathematics, Northwestern University, 2033 Sheridan Road, Evanston, IL 60208-2730; email: `priddy@math.nwu.edu`

Combinatorics and Lie theory to SERGEY FOMIN, Department of Mathematics, University of Michigan, Ann Arbor, Michigan 48109-1109; email: `fomin@umich.edu`

Complex analysis and complex geometry to DUONG H. PHONG, Department of Mathematics, Columbia University, 2990 Broadway, New York, NY 10027-0029; email: `phong@math.columbia.edu`

*****Differential geometry and global analysis** to LISA C. JEFFREY, Department of Mathematics, University of Toronto, 100 St. George St., Toronto, ON Canada M5S 3G3; email: `jeffrey@math.toronto.edu`

Dynamical systems and ergodic theory to ROBERT F. WILLIAMS, Department of Mathematics, University of Texas, Austin, Texas 78712-1082; email: `bob@math.utexas.edu`

*****Geometric analysis** to TOBIAS COLDING, Courant Institute, New York University, 251 Mercer Street, New York, NY 10012; email: `colding@cims.nyu.edu`

Geometric topology, knot theory and hyperbolic geometry to ABIGAIL A. THOMPSON, Department of Mathematics, University of California, Davis, Davis, CA 95616-5224; email: `thompson@math.ucdavis.edu`

Harmonic analysis, representation theory, and Lie theory to ROBERT J. STANTON, Department of Mathematics, The Ohio State University, 231 West 18th Avenue, Columbus, OH 43210-1174; email: `stanton@math.ohio-state.edu`

*****Logic** to THEODORE SLAMAN, Department of Mathematics, University of California, Berkeley, CA 94720-3840; email: `slaman@math.berkeley.edu`

Number theory to HAROLD G. DIAMOND, Department of Mathematics, University of Illinois, 1409 W. Green St., Urbana, IL 61801-2917; email: `diamond@math.uiuc.edu`

*****Ordinary differential equations, and applied mathematics** to PETER W. BATES, Department of Mathematics, Michigan State University, East Lansing, MI 48824-1027; email: `peter@math.msu.edu`

*****Partial differential equations** to PATRICIA E. BAUMAN, Department of Mathematics, Purdue University, West Lafayette, IN 47907-1395' email: `bauman@math.purdue.edu`

*****Probability and statistics** to KRZYSZTOF BURDZY, Department of Mathematics, University of Washington, Box 354350, Seattle, Washington 98195-4350; email: `burdzy@math.washington.edu`

Real analysis and partial differential equations to DANIEL TATARU, Department of Mathematics, University of California, Berkeley, Berkeley, CA 94720; email: `tataru@math.berkeley.edu`

All other communications to the editors should be addressed to the Managing Editor, WILLIAM BECKNER, Department of Mathematics, University of Texas, Austin, TX 78712-1082; email: `beckner@math.utexas.edu`.

Titles in This Series

774 **Kai A. Behrend,** Derived ℓ-adic categories for algebraic stacks, 2003

773 **Robert M. Guralnick, Peter Müller, and Jan Saxl,** The rational function analogue of a question of Schur and exceptionality of permutation representations, 2003

772 **Katrina Barron,** The moduli space of $N = 1$ superspheres with tubes and the sewing operation, 2003

771 **Shigenori Matsumoto,** Affine flows on 3-manifolds, 2003

770 **W. N. Everitt and L. Markus,** Elliptic partial differential operators and symplectic algebra, 2003

769 **Jie Wu,** Homotopy theory of the suspensions of the projective plane, 2003

768 **R. Höpfner and E. Löcherbach,** Limit theorems for null recurrent Markov processes, 2003

767 **Po Hu,** S-modules in the category of schemes, 2003

766 **Su Gao and Alexander S. Kechris,** On the classification of Polish metric spaces up to isometry, 2003

765 **Robert Bieri and Ross Geoghegan,** Connectivity properties of group actions on non-positively curved spaces, 2003

764 **J. Spandaw,** Noether-Lefschetz problems for degeneracy loci, 2003

763 **Yasuyuki Kachi and Eiichi Sato,** Segre's reflexivity and an inductive characterization os hyperquadrics, 2002

762 **Leiba Rodman, Ilya M. Spitkovsky, and Hugo Woerdeman,** Abstract band method via factorization, positive and band extensions of multivariable almost periodic matrix functions, and spectral estimation, 2002

761 **Oliver Druet and Emmanuel Hebey,** The AB program in geometric analysis : Sharp Sobolev inequalities and related problems, 2002

760 **Markus Banagl,** Extending intersection homology type invarients to non-Witt spaces, 2002

759 **Donald M. Davis,** From representation theory to homotopy groups, 2002

758 **Alan Forrest, John Hunton, and Johannes Kellendonk,** Topological invariants for projection method patterns, 2002

757 **Douglas Bowman,** q-difference operators, orthogonal polynomials, and symmetric expansions, 2002

756 **José Ignacio Cogolludo-Agustín,** Topological invariants of the complement to arrangements of rational plane curves, 2002

755 **M. A. Mandell and J. P. May,** Equivariant orthogonal spectra and S-modules, 2002

754 **Edward L. Green, Idun Reiten, and Øyvind Solberg,** Dualities on generalized Koszul algebras, 2002

753 **Daniel Panazzolo,** Desingularization of nilpotent singularities in families of planar vector fields, 2002

752 **Linus Kramer,** Homogeneous spaces, Tits buildings, and isoparametric hypersurfaces, 2002

751 **Bruce Allison, Georgia Benkart, and Yun Gao,** Lie algebras graded by the root systems BC_r, $r \geq 2$, 2002

750 **Masaki Izumi and Hideki Kosaki,** Kac algebras arising from composition of subfactors: General theory and classification, 2002

749 **Nanhua Xi,** The based ring of two-sided cells of affine Weyl groups of type $\widetilde{A}_{n-1}$, 2002

748 **Jürgen Ritter and Alfred Weiss,** The lifted root number conjecture and Iwasawa theory, 2002

747 **Armand Borel, Robert Friedman, and John W. Morgan,** Almost commuting elements in compact Lie groups, 2002

TITLES IN THIS SERIES

746 **Peter Niemann,** Some generalized Kac-Moody algebras with known root multiplicities, 2002

745 **Mikhail A. Lifshits and Werner Linde,** Approximation and entropy numbers of Volterra operators with application to Brownian motion, 2002

744 **Roger Chalkley,** Basic global relative invariants for homogeneous linear differential equations, 2002

743 **Heng Sun,** Spectral decomposition of a covering of $GL(r)$: the Borel case, 2002

742 **J. E. Gilbert, Y. S. Han, J. A. Hogan, J. D. Lakey, D. Weiland, and G. Weiss,** Smooth molecular functions and singular integral operators, 2002

741 **Francisco Santos,** Triangulations of oriented matroids, 2002

740 **Rick Durrett,** Mutual invadability implies coexistence in spatial models, 2002

739 **Georgios K. Alexopoulos,** Sub-Laplacians with drift on Lie groups of polynomial volume growth, 2002

738 **Yasuro Gon,** Generalized Whittaker functions on $SU(2,2)$ with respect to the Siegel parabolic subgroup, 2002

737 **Arjen Doelman, Robert A. Gardner, and Tasso J. Kaper,** A stability index analysis of 1-D patterns of the Gray-Scott model, 2002

736 **Wojciech Chachólski and Jérôme Scherer,** Homotopy theory of diagrams, 2002

735 **Martina Brück, Xi Du, Joonsang Park, and Chuu-Lian Terng,** The submanifold geometries associated to Grassmannian systems, 2002

734 **Michel Van den Bergh,** Blowing up of non-commutative smooth surfaces, 2001

733 **Milé Krajčevski,** Tilings of the plane, hyperbolic groups and small cancellation conditions, 2001

732 **Jan O. Kleppe, Juan C. Migliore, Rosa Miró-Roig, Uwe Nagel, and Chris Peterson,** Gorenstein liaison, complete intersection liaison invariants and unobstructedness, 2001

731 **Jesús Bastero, Mario Milman, and Francisco J. Ruiz,** On the connection between weighted norm inequalities, commutators and real interpolation, 2001

730 **Suhyoung Choi,** The decomposition and classification of radiant affine 3-manifolds, 2001

729 **Michael Grosser, Eva Farkas, Michael Kunzinger, and Roland Steinbauer,** On the foundations of nonlinear generalized functions I and II, 2001

728 **Laura Smithies,** Equivariant analytic localization of group representations, 2001

727 **Anthony D. Blaom,** A geometric setting for Hamiltonian perturbation theory, 2001

726 **Victor L. Shapiro,** Singular quasilinearity and higher eigenvalues, 2001

725 **Jean-Pierre Rosay and Edgar Lee Stout,** Strong boundary values, analytic functionals, and nonlinear Paley-Wiener theory, 2001

724 **Lisa Carbone,** Non-uniform lattices on uniform trees, 2001

723 **Deborah M. King and John B. Strantzen,** Maximum entropy of cycles of even period, 2001

722 **Hernán Cendra, Jerrold E. Marsden, and Tudor S. Ratiu,** Lagrangian reduction by stages, 2001

721 **Ingrid C. Bauer,** Surfaces with $K^2 = 7$ and $p_g = 4$, 2001

720 **Palle E. T. Jorgensen,** Ruelle operators: Functions which are harmonic with respect to a transfer operator, 2001

719 **Steve Hofmann and John L. Lewis,** The Dirichlet problem for parabolic operators with singular drift terms, 2001

For a complete list of titles in this series, visit the AMS Bookstore at **www.ams.org/bookstore/**.